To Scott & ↑

P. 147 was just
for you.

Chris

Passive Triggering of a High-Power
Hollow-Cathode
EUV Lamp for Lithography

Von der Fakultät für Maschinenwesen der Rheinisch-Westfälischen Technischen Hochschule Aachen zur Erlangung des akademischen Grades eines Doktors der Naturwissenschaften genehmigte Dissertation

Vorgelegt von

Christopher Sydney Smith M.Sc.

aus Sheffield, England

Referent: Univ.-Prof. Dr. rer. nat. Reinhart Poprawe M.A.
Korreferent: Univ.-Prof. Dr. rer. nat. Kurt Kugeler

Tag der mündlichen Prüfung: 16. August 2005
D 82 (Diss. RWTH Aachen)

Berichte aus der Physik

Christopher Sydney Smith

Passive Triggering of a High-Power Hollow-Cathode EUV Lamp for Lithography

D 82 (Diss. RWTH Aachen)

Shaker Verlag
Aachen 2006

Bibliographic information published by Die Deutsche Bibliothek
Die Deutsche Bibliothek lists this publication in the Deutsche
Nationalbibliografie; detailed bibliographic data is available in
the internet at http://dnb.ddb.de.

Zugl.: Aachen, Techn. Hochsch., Diss., 2005

Printed in Germany.

ISBN 3-8322-4910-9
ISSN 0945-0963

Shaker Verlag GmbH • P.O. BOX 101818 • D-52018 Aachen
Phone: 0049/2407/9596-0 • Telefax: 0049/2407/9596-9
Internet: www.shaker.de • eMail: info@shaker.de

KURZFASSUNG

Die Lichtquelle ist die kritischste Komponente für die Realisierung extrem ultravioletter (EUV) Lithographie und der Produktion immer kleineren Mikroprozessoren. Um bis 120 Siliziumwafer pro Stunde in einem kommerziellen Scanner zu verarbeiten, wäre 115 W gesammelte Inband-EUV Strahlung nötig.[1] Diese Anforderungen an eine kommerzielle EUV Quelle für Massenproduktion der Wafern übersteigen noch die gegenwärtige Leistung durch große Faktoren, sowohl in Ausgangsenergie als auch in Lebenszeit.

Die hohlkathodegetriggerte (HCT) Lampe wurde gewählt für die Weiterentwickelung über den gegenwärtigen Stand der Technik hinaus, zur höheren Leistung und Stabilität. Bisherige Arbeiten über die HCT Lampe benutzen ein selbsttriggerndes System, welche limitiert ist durch seine Stabilität. Neue Fortschritte werden für die Entwicklung der EUV Quelle geschafft, indem man die Gasentladung auslöst (Triggerung). Der Entladungsprozess kann durch das Einbringen einer Triggerelektrode im System groß beeinflusst werden. Der Betrieb, der nur eine Blockierelektrode an einem positiven Potential relativ zur Kathode verwendet, wird erfolgreich benutzt, um die Lampe mit Wiederholungsraten von bis zu 4,5 kHz mit reproduzierbaren Zündverzögerungszeiten von einigen μs zu betreiben. Andere Triggerkonzepte werden auch untersucht, wobei Betrieb bis zu 7 kHz erreicht wird, welche 10,2 W von 13,5 nm Strahlung produziert.

Der Vorteil des getriggerten Betriebs über ungetriggerten Betrieb wird erkannt durch das Verbreitern des Parameterraums, eine exaktes Timing des Pulses und die möglichere Veränderung des Betriebsdrucks und der Spannung unabhängig von einander.

ABSTRACT

The light source is the most critical component for the realisation of extreme ultraviolet (EUV) lithography and the production of ever smaller microchips. In order to process up to 120 silicon wafers per hour in a commercial scanner, 115 W of collected in-band EUV radiation would be required.[1] These requirements on a commercial EUV source for volume production of wafers still exceed the current performance by large factors, both in terms of output power and lifetime.

The hollow-cathode triggered (HCT) lamp is chosen for scaling-up beyond the current best available technology, to a higher power and stability. Previous works on the HCT lamp make use of a self-triggering system, which is limited by its stability. New progress is made on the development of the EUV source by triggering the gas discharge. The breakdown process can be largely influenced by introducing a trigger electrode to the system. Operation using only a blocking electrode at a positive potential with respect to the cathode is used to trigger the lamp at repetition rates of up to 4·5 kHz with a reproducible ignition delay time of a few μs. Other triggering concepts are also investigated allowing operation of up to 7 kHz producing 10·2 W of 13·5 nm radiation.

The advantage of triggered operation over untriggered operation is found in the broadening of the parameter space and allows precise timing of the pulses and variation of the operating pressure and voltage independently of one another.

Contents

1. Introduction

1.1. Microprocessor Production

Microprocessor chips belong to the most complex items ever manufactured and require hundreds of steps in the cleanest of environments. Silicon wafers, cut from pure silicon, are the starting point for their manufacture. Metals, such as aluminium and copper, are used as conductors, and gold is used to connect the chips to their packaging.

Ultraviolet light is used to expose patterns on the silicon in a process called lithography. Masks are used in the chip-making process like stencils. Microprocessors are built in layers on a silicon wafer through various processes using chemicals, gases and UV light. On the wafer, the first layer of silicon dioxide is grown by heating the silicon wafer to a very high

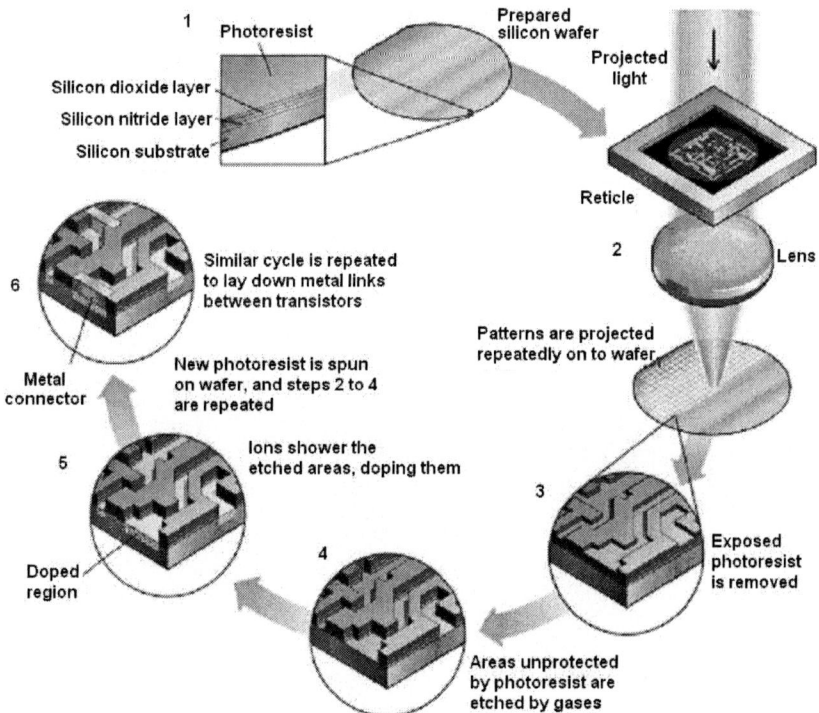

Figure 1.1: Schematic of the microprocessor production.

temperature in a special furnace in the presence of oxygen. Silicon dioxide has two important properties, firstly it is a high quality electrical insulator, and secondly it acts as a barrier to impurities. The wafer is then coated with a substance called the photo-resist.

During the photolithography stage, UV light passes through the mask, onto the wafer. The mask protects parts of the wafer from the light and the unprotected areas of the photo-resist become soluble when exposed to the light. The photo-resist is then completely dissolved by a solvent, leaving only a pattern of the photo-resist on the silicon dioxide. The exposed silicon dioxide is etched away chemically. The rest of the photo-resist is removed and ridges of silicon dioxide are left on the silicon wafer. An ion implantation process is then used to dope the exposed areas of the silicon wafer, forming impurities in the silicon to vary its resistivity. Ions are implanted in the wafer to alter the semiconductor properties of the silicon in these areas.

The chips are built up in layers, so the processes are repeated, a new layer of silicon dioxide is grown over the ridges and etched areas of the wafer. A layer of polysilicon and another layer of photo-resist are then applied. UV light is then passed through a second mask, exposing a new pattern on the photo-resist. This is again dissolved in solvent exposing the polysilicon and silicon dioxide, which are again etched away by chemicals.

Figure 1.2: An ASML 193 nm Stepper/Scanner used for current chip production.

The layering and masking processes are repeated, creating gaps that allow for connections to be made between the layers. Atoms of metal are deposited on the wafer, filling the gaps. Another masking and etching stage leaves strips of the metal that make electrical connections.

Roughly 20 layers are connected to form the microprocessor's circuitry in a 3-D structure. The exact number of layers depends on the particular design. In reality more than 250 steps are required and hundreds of identical microprocessors are created in batches on each wafer, which is cut with a diamond saw to separate the chips.

1.2. Motivation

In 1965 Gordon Moore made the, now famous, observation that there was an exponential growth in the number of transistors in an integrated circuit.[2] The number of transistors has doubled every eighteen months, and still continues to do so.

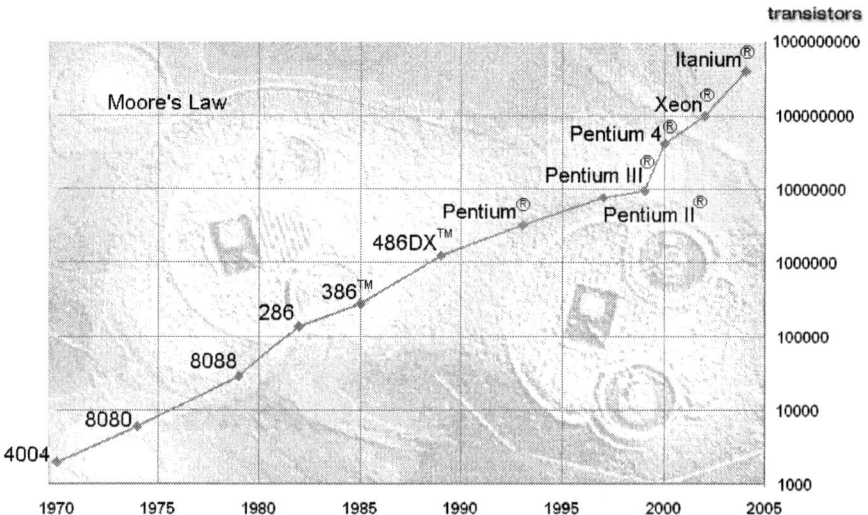

Figure 1.3:Moore's law states that the rate of growth in the number of transistors on an integrated circuit has doubled approximately every 18 months.

If Moore's law is to remain true the chips will soon have to become larger or the size of the components on the chips reduced. There are however physical limits to both. Doubling the size of the chip every eighteen months would soon lead to problems of the physical size, economy and also heat removal from the components.[3] In the opposite direction, as the size of the components become close to the wavelength of the light used to create them, it becomes increasingly difficult to resolve the structures. The resolution of an optical system is defined as

$$RES \propto \frac{\lambda}{NA} \qquad (1.1)$$

where λ is the wavelength of the light and NA the numerical aperture.[4] The numerical aperture of a system is defined by the maximum angle, θ from which light can enter the optics

$$NA = n \cdot \sin(\theta) \qquad (1.2)$$

where n is the refractive index. As the equation shows a better resolution is achieved by reducing the wavelength or increasing the numerical aperture. This has however an effect on the depth of focus

$$DOF \propto \frac{\lambda}{(NA)^2} \qquad (1.3)$$

making the more favourable solution therefore a light source with a shorter wavelength.

The wavelength of 13·5 nm was chosen for the next generation lithography. Multi-layer reflective coatings can be used to make up multi-component optical systems for a wide range of wavelengths, achieved by applying multi-layer coatings (e.g. Molybdenum-Silicon or Molybdenum-Beryllium) to a surface. The coatings are described as artificial Bragg crystals; a high atomic number material takes the place of the atom planes and a low atomic number, the material separating them.[6]

The efficiencies of such mirrors, however, also vary widely depending on the wavelength and which materials are used. Mo-Si and Mo-Be multi-layers that peak at 13·5 nm have achieved reflectivities exceeding of 68%, making them the highest near-normal-incidence reflectivity of any multi-layer mirrors in the EUV range.[59, 60]

The requirements on a source are derived from an economic throughput model for the processing of up to 120 wafers/h in a commercial EUV scanner. With the current design of scanner, this means a requirement of around 120 W. The 120 W refers to the collected "in-band" EUV power emitted within 2% of the centre wavelength of 13·5 nm. For etendue[*] limits imposed by the optical system, the radiation must be emitted out of a small volume of several mm^3 in order to be collectable by the optical system. Typical collector openings are limited to a geometrical opening of 2 - 3 sr, whilst collector efficiencies are in the order of 50 % - 80 %.

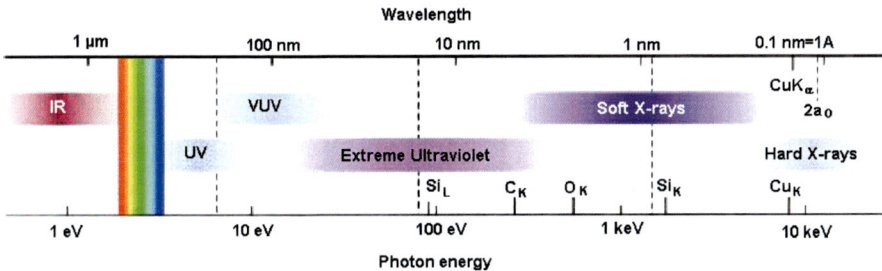

Figure 1.4: Spectrum showing the EUV region.

Today's requirements on a commercial EUV source for high-volume production of wafers still exceed the current performance by large factors both in terms of output power and lifetime. The light source has been identified as being the most critical single component on the way to realising EUV lithography. It is well known that hot dense plasmas are a source of radiation in the EUV region.[5] For this reason much research is taking place on plasma sources. The energy region (around 100 eV) is also of interest for X-ray optics based on Si-Mo multi-layer mirrors.

*) The étendue, E is a characteristic of an optical instrument. It is a measure of the light gathering power, i.e. the power transmitted per radiance of the source. If A is the area of the source and Ω the solid angle accepted from each point of the source by the aperture then
$$E = A\Omega = \Phi/L \ [m^2 \ sr].$$

1.3. Current Technology for EUV Radiation

1.3.1. Plasma Sources - General

Both Laser Produced Plasma (LPP) and Discharge Produced Plasma (DPP) concepts are being studied extensively for the so-called "Next Generation Lithography". The principle of a plasma source follows nothing other than Wien's law,

$$\lambda_{max} \cdot T = \text{constant,} \qquad (1.4)$$

where λ_{max} defines the wavelength with the greatest amplitude emitted from a body at a temperature, T. So, if an object is hot enough it will radiate UV or X-radiation in the form of a spectrum based upon the blackbody spectrum for the temperature of the plasma.[7]

However at the temperatures required for 13 nm radiation (~20 eV), the electrons and ions will separate and a plasma will form. There are several methods of generating hot plasmas which radiate in this spectral region. Currently under investigation are: Laser Produced Plasma, Plasma Focus, Capillary Discharge, Z-pinch, Star Pinch and the Hollow Cathode Triggered pinch on which this work is based.

1.3.2. Laser Produced Plasma Sources

When an intense laser beam is focused on the surface of a target material, usually under vacuum conditions, plasma is created. Due to the high power of the laser pulse many of the target atoms are ionised; the level of ionisation being dependent on the temperature of the plasma created.

As the ions decay from their excited states they emit characteristic radiation and during the lifetime of the laser pulse thermal equilibrium can be achieved at the required high temperature giving a broadband emission. Figure 1.5 shows a schematic of such a system.

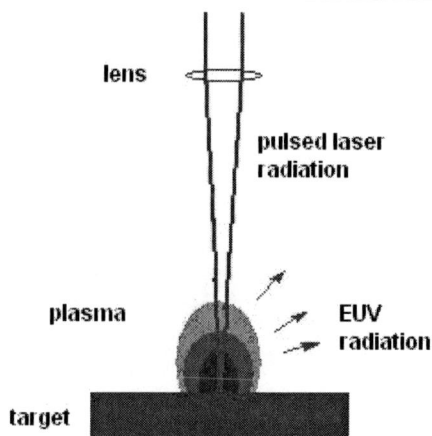

Figure 1.5: Laser produced plasma principle.[8]

From a low atomic mass element, such as lithium, characteristic line emissions are observed, and from high atomic mass elements, such as xenon, a broad band spectrum is seen.[9] Laser sources, should in principle be able to run at the high repetition rates required for a commercial source, although much debris caused by the beam on the target is left in the system. This problem also arises for a laser-triggered gas discharge (see 3.2.3). The target does not necessarily have to be a solid. Using cryogenically frozen xenon as a target, the company Power Laser claim a conversion efficiency of 0·9 % using a 600 W solid state laser. This produces a yield of about 5·4 W/2π.sr of EUV radiation (2·7 W at the 2nd focus).

1.3.3. Capillary Discharge Sources

A capillary discharge source consists of a tube (the capillary) made for instance, from lithium or aluminium oxides, to which a very large current can be applied along its length, see figure 1.6. Like all discharge sources with a current in the kA range, and also for thermal reasons, the current is applied in pulses. An atmosphere of oxygen or argon at around 1-10 Pa is often used to vary the properties of the pinch. When the large current is present on the inner surface of the tube a plasma begins to form from the tube material.

This plasma then pinches under the Lorentz force towards the axis of the capillary (see 2.3.4 on Pinch Formation).

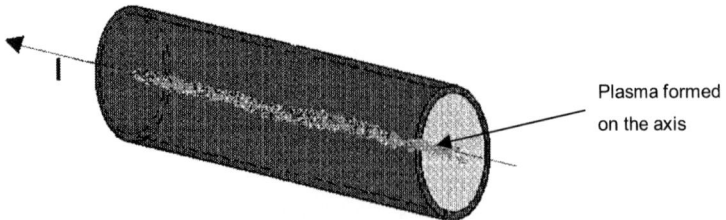

Plasma formed
on the axis

Figure 1.6:A capillary discharge source, plasma is formed from the wall material and pinches to the axis under the large current.

This hot dense plasma is then capable of emitting radiation in the extreme ultraviolet region. As the plasma pinches more of the surface material is eroded. The shock on the material caused by the pulsed contact with the plasma limits the lifetime of such a device to only several thousand pulses, making high repetition rates impossible.

1.3.4. Z-Pinch

The z-pinch is a discharge device consisting of two electrodes separated by an insulating tube. Z-pinch systems use large electric currents to produce a plasma from a fill gas or by vaporising very thin wires or foils on the axis. The high currents produce powerful magnetic fields that surround the plasma, pinching it on the axis, to densities and temperatures sufficient to generate an intense source of x-rays. When the confined plasma "pinches" on axis, the energy stored in the plasma is extracted as a short burst of intense x-rays, before the instabilities can destroy the symmetric geometry. This is not a new technology: The concept was studied experimentally in the 1950's as a way to achieve fusion conditions in the laboratory using deuterium, but it was not possible to scale-up such systems to the required power.

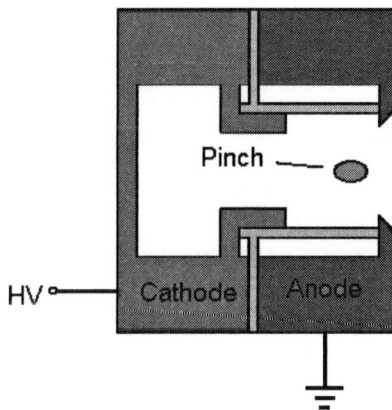

Figure 1.7: Schematic of a z-pinch. The main discharge may be initiated by a surface discharge or other trigger mechanism at the rear of the system.

A puff of gas may also be introduced between two electrodes; when voltage is applied, a current flows through the gas heating and ionising it. The current produces again a powerful magnetic field that surrounds the ionised material (or plasma) and accelerates it radially inward toward its axis (the z-axis, hence the name "z-pinch").

Traditionally the z-pinch system suffers from a very short lifetime, as the insulator becomes metallised. The company Xtreme Technologies, are working on such a source and claim to have a lifetime now of 10^8 shots by using a large separation between the electrodes and the plasma, with repetition rates up to 4 kHz in burst operation delivering up to 60 W/2π.sr (corresponding to 5W in the 2nd focus).[10]

1.3.5. Plasma Focus

Like the z-pinch, the plasma focus is a device which can discharge currents from a few kA to several MA and voltages of several tens of kV. It was also developed for fusion experiments. Such a device is shown in figure 1.8. The system has two cylindrical coaxial electrodes placed in a vacuum chamber. The two electrodes are separated at one end by an insulating sleeve (generally heat-proof glass) while the other end is left free. Gas is let into the discharge chamber at pressures of a few mbar.[11]

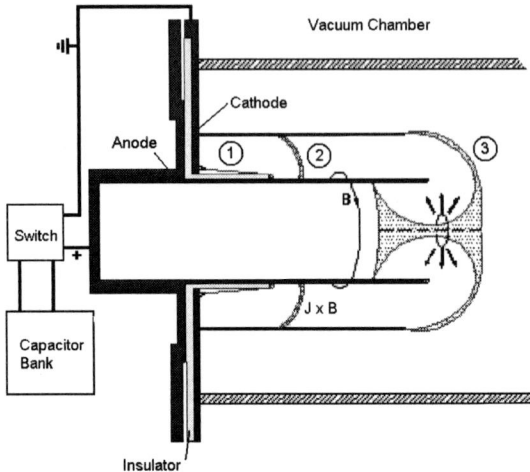

Figure 1.8: Schematic of a plasma focus device showing the phases of its function.[11]

After a capacitor bank has been charged up a fast switch for high currents (e.g. a spark-gap) is closed the discharge develops in four distinct phases.

Breakdown: When the switch is closed, a thick plasma current sheath is generated on the insulator which separates the rear end of the two electrodes. As the current increases, the sheath "explodes" in a radial direction driven by the radial component of the Lorentz force.

Acceleration: Driven by the axial component of the force, the sheath moves rapidly towards the front end of the electrodes and in doing so, the sheath collects all the gas present between electrodes in front of it ("snowplough" effect). On reaching the front of the electrodes it encircles the inner electrode.

Compression: This critical phase of the discharge begins after the inner electrode has been encircled and ends with the formation of hot, dense, focalised plasma. The sheath collapse and constriction mechanism are explained as a "pinch" effect as before. It is confined along the symmetry axis of the electrodes, exactly in front of the central electrode, in a small cylindrical zone. The open source geometry of the plasma focus offers a high collection angle for radiation. To maximise the efficiency of the plasma heating process, the current that flows in the circuit must reach its peak at the start of compression.

Expansion: Once maximum constriction of the sheath is reached, instabilities occur in the plasma, destroying the pinch effect and causing the column to expand. (This is not shown in the figure).

The performance of such a plasma focus source is stated as 2 kHz delivering 66 W/2π sr (CW) and 5 kHz delivering 200 W/2π sr (burst) by the company Cymer.[12]

1.3.6. Star Pinch

In a similar construction to the 12-channel pseudospark switch in (13), multiple electrodes outside a large plasma chamber are used to project small xenon beams. These beams, from both above and below, meet in the centre where they are neutralised by charge exchange. A voltage then applied between the main upper and lower electrodes initiates a high current which heats the plasma and leads to a pinch emitting EUV. Such a source, known as an Astron or Star Pinch, is being pursued by the company Plex LLC[14], who claim to have EUV powers of up to 6 W at the source and repetition rates of up to 700 Hz. The central insulator is protected by the magnetic field of the pinch, keeping the plasma far enough away (>10 mm) as not to cause erosion or metallisation.[14]

1.3.7. Synchrotron Radiation

EUV radiation can be generated using a synchrotron (a torus-shaped particle accelerator). Bremsstrahlung radiation associated with the acceleration of the electrons towards the centre of the circle is emitted as a characteristic energy corresponding to the radius of the circle. The rings are usually quite large in diameter, up to many tens of metres.

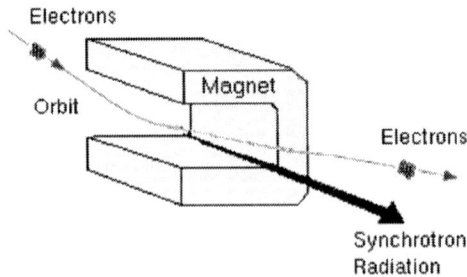

Figure 1.9: Part of a synchrotron radiation source.

The radiation spectrum from a traditional synchrotron is a simple broadband spectrum, produced typically having a small distribution of energies representative of the particle speed distribution. The circular speed represents the temperature of the plasma in the ring and in turn the energy of the emission (see also 1.4.1). For non-relativistic motion, the radiation spectrum is simple and is called "cyclotron radiation". The frequency of radiation is simply the gyration frequency, which is given in terms of the magnetic field as

$$v = \frac{eB}{mc} \qquad (1.5)$$

where B is the field strength, e is the electric charge, m the electronic mass, and c the speed of light. Clearly, this synchrotron emission applies only to particle motion perpendicular to the direction of a magnetic field; real particles also have particle motions parallel to the field (the thickness of the beam), and radiate thermally from this component of their motion.

Using a set of alternate magnetic poles in the beam line, known as an undulator or wiggler, this component can be enhanced greatly to produce (almost) monochromatic radiation. In the centre-of-mass frame of reference (travelling with the speed of the beam) the radiation *is* monochromatic, with the possibility of higher harmonics of oscillation being present. When transferred to the laboratory frame (stationary) the peaks are broadened due to a Doppler shift caused by distribution of directions in which the radiation can be emitted. When more magnets are used the peaks are reduced in width.

1.3.8. The Hollow-Cathode Triggered Discharge

The remaining source is the HCT source, and is the source chosen for this work. It is also a kind of Z-pinch discharge, but differs in the fact that it is based on a pseudospark geometry, allowing self-ignition and less current contact with the electrodes. Such sources are already commercially available. The company AixUV produces low-power lamps based on this method allowing powers of up to 150 mW/2π.sr and repetition rates of up to 250 Hz. The function of this lamp is discussed in detail in the following sections.

1.4. Source Selection and Objectives

The seven sources presented all have ability to generate a high temperature plasma, radiating in the EUV region. The hollow-cathode device, which is considered in this work, allows the generation of a plasma overcoming the known problems, which the other six possess. Gas discharges should generally have fewer problems with debris over lasers or capillaries, as the gas can be pumped away and no solid material has to be used up. The insulator is hidden away from the plasma, whereas the standard z-pinch suffers from metallisation of the open insulator. Its electrodes also suffer erosion from the current contact. The special choice of electrodes, which are based on pseudospark discharge geometry allow very large currents with a much lower current density. The plasma-focus and the star pinch both have large distances between the electrodes and the plasma, which may be a thermal advantage, yet they may not be able to reach the high frequencies required, due to the slower temporal development. Synchrotrons are simply too large and very expensive. A summary of the best available technology for producing 13·5 nm radiation is presented in figure 1.10. In general, there are only low power sources are available, with a wide range of lifetimes, whereas the increased power sources show an exponential decay to their lifetime with the output power. Differing technologies are difficult to compare, and for this reason, the lifetime here means the time when the source can no longer be used; be it due to debris on the optics or source of EUV itself being damaged.

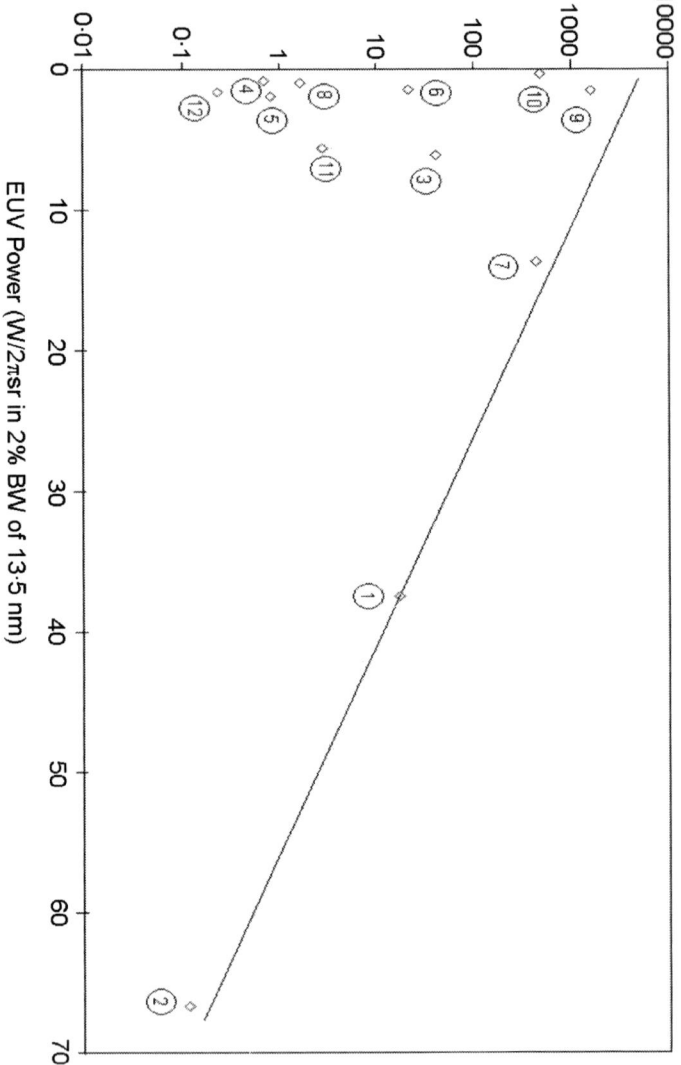

Figure 1.10: Parameter range, showing source lifetime (or attatched optics, whichever is shorter) against output power for currently available sources of 13·5 nm radiation. The data are taken from proceedings of the EUVL Source Workshop, Santa Clara, USA, 23.02.2003. The state of the art technology shows an exponential relation between output power and source lifetime.

To create a successful EUV lamp suitable for commercial use this trend must be broken; both high power and high life time are required.

A low-power HCT lamp has already relatively good lifetime characteristics, ⑩ in figure 1.10. When increased in repetition rate a considerable loss occurs, see ⑪ in the same figure, yet this loss tends to be due to the wrong materials and design (the original lamp was only intended for low power, and the only barrier to reaching higher power lies with stability of the parameters of the lamp, and their dependence on each other. Hence, for stable operation of a pulsed source above about one kilohertz, triggering concepts are necessary to replace the self-discharging system used up until now and to break these inter-parameter dependencies.

Through investigation into the operation of the lamp, the objective is to find a trigger, whose operation should widen the available parameter range of variables, such as pressure and voltage, for the discharge, whereby the possibilities for optimisation of the pulse output increase, as well as reducing the jitter inherent to the operation of the current self-discharging system. The greatest meaning then for the introduction of the trigger, is that the pressure may be chosen by the operator of the lamp, rather than being determined by the voltage or how long this has been present on the system. Similarly, the operator will be able to vary the voltage at which the lamp runs, whilst keeping the pressure constant. And, not insignificantly, he may choose *when* the lamp discharges, rather than being subject to random processes.

The electrode system must be modified and fitted out with cooling in preparation for operation at higher powers. New materials for the electrodes need to be tried and tested so that the EUV-source can be successfully integrated into a commercial lithography system. Particular attention must also be paid to the long-term stability of the source, not just the lifetime. As the electrodes are subject to an intense thermal load, they will also be eroded over time.

The overall objective is to bring the gas discharge source, based on a hollow-cathode triggered discharge, up to a useful radiation level of around 50 W in the range around 13·5 nm. Experimental proof is also to be provided that the scaling up to a light power necessary for a production source can be achieved.

2. Theory

2.1. Electrical Discharges

Before the HCT lamp can be discussed, some basic theory[15] of electrical discharges should be understood. If two electrodes in a gas-filled system are given a potential difference, V, and are separated by a distance, d, then it is fairy well known that if V is large, or d is small, a discharge can occur. Similarly if the pressure, p, is reduced the likelihood of a discharge is increased.

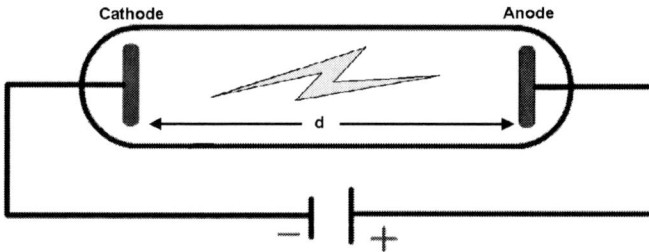

Figure 2.1: Electronic schematic of a gas discharge system.

Let us consider in more detail the moment instantaneously after the voltage has been applied to the electrode pair. No current is flowing and the voltage remains at V. If however, free electrons or ions are present in the gap between the electrodes; these will be accelerated by the electric field in the direction of the anode or cathode respectively. With the acceleration before collision of

$$a = eV / m_e d \qquad (2.1)$$

where e and m_e are the charge and the mass of the electron (or ion) respectively. These free charge carriers may come from many sources, for instance, cosmic rays, radioactive decay, photoelectric emission (if light is present on the surface) or emission directly from the cathode surface if it is hot. The source is unimportant at the moment, but will be a basis for discussion when wanting to trigger or prevent a discharge. At this early stage of discharge formation a current is already flowing and the potential difference between the electrodes reduces. The current, caused by the drift of the electrons, is represented by

$$i_e = n_e e v_d A \qquad (2.2)$$

where n_e is the number of electrons, v_d, their average drift velocity (including collisions) and A the area through which the current is flowing. A similar term should be added for the ions, although initially the electrons move much faster due to their lighter mass. The reduction in voltage is caused by the negative electrons moving closer to the anode and positive ions to the cathode.

If after an electron has moved some distance, λ, a collision takes place with one of the neutral gas atoms (or molecules); one of several possible processes can take place. The electron may be reabsorbed (killing off the discharge process), it may be scattered into a different direction (elastically or inelastically), after which it will again be subject to the acceleration of the field, or, most importantly, it may knock a second electron out of its orbit around the atom, leaving two electrons subject to the field. Whether or not a particular atom will be ionised depends on how much kinetic energy the electron has when it collides. This process can be repeated many times in steps of the distance, λ, until the electrons reach the anode. The distance λ is defined as the mean free path of an electron in a gas with N molecules per unit volume and an interaction cross-section, σ.

$$\lambda = 1 / N\sigma \qquad (2.3)$$

The cross-section represents the probability that an interaction will take place, which represents the statistical nature of the processes on the atomic level. The value of λ in air at standard temperature and pressure is around 100 nm; so many electrons can be released, each time increasing the current exponentially. If the energy of the incoming electron is greater than the binding energy of the electrons in the atom (the "ionisation potential") then one or more electrons can be released. Unlike the atoms of the gas, whose kinetic energy depends on the gas temperature ($E = {}^3/_2.kT$), the kinetic energy of the electrons is determined by the electric field and the distance over which this accelerating force acts.

This distance that the electron can accelerate is, however, limited to the mean free path in the gas. Each time the electron collides with an atom, it looses most of its kinetic energy. If the distance the electron accelerates through is very small (i.e. at high pressure), it will never have enough kinetic energy to ionise the gas. This is why lowering the pressure requires a lower voltage. Longer gaps reduce the electric field strength and consequently reduce the kinetic energy the electrons can obtain.

This process of multiplication is known as a Townsend avalanche and leaves behind a quasi-neutral cone of plasma. Quasi-neutral meaning that the electrons are separated from the ions, yet on the whole there is no net charge. Obviously more than one avalanche can take place at the same time, but in theory one electron is enough to achieve discharge in a gap. The cone (sometimes referred to as a teardrop) of plasma will start to spread out due to like charges repelling and the force from the electric field will pull the ions towards the cathode and the electrons (much faster due to their lower mass) to the anode.

To many people the term "discharge" means a short circuit between the electrodes. This does not have to be the case. As seen above, the flow of only a small number of electrons/ions corresponds to a current. The resistivity of the gap is still high and it may well be that a discharge never develops beyond this point. Assuming however that it does, the space charge of the plasma left by the avalanche spreading out will cause a distortion of the electric field, which is shown schematically in figure 2.2. In front and behind the quasi-neutral zone, new avalanches can form in the now increased field. These may be initiated by ions striking the cathode releasing secondary electrons or maybe light emitted by the plasma due to the photoelectric effect. These new avalanches are directed towards the central expanding channel due to the distorted field. The resistivity of the central channel reduces as more and more ionisation occurs. The electrons will gradually leave the region as they accelerate faster than the ions and the volume starts to become positively charged; the electric field distorts further. It is the geometric influences on the development of this dynamic process which are crucial to the operation of the HCT lamp.

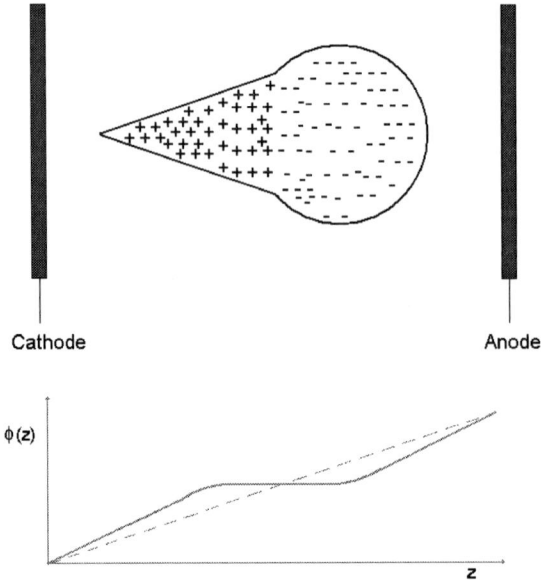

Figure 2.2: Above: Schematic of the space-charge created by the Townsend avalanche.
Below: Potential distribution across the gap at this point. The dashed line shows
the original field before the avalanche took place.

Plasma is conducting, and dense plasma even more so, so inside the plasma (as in a conductor) there can be little field. This effect, called Debye shielding, "squashes up" the applied electric field outside the plasma, nearer to the walls of the system, where electrons can be acquired or given up readily. In other words, plasma cannot be in contact with the walls of the system as it becomes neutralised. The distance over which the electric field dies away is called the "sheath" and is described by a few times the Debye length, λ_D [58]

$$\lambda_D = \frac{1}{e}\sqrt{\frac{\varepsilon_o kT}{n_e}} \qquad (2.4)$$

where ε_o is the permittivity of free space, k = Boltzmann's constant and T the temperature of the gas (plasma). Obviously the plasma cannot suddenly exist at a certain point, but rather grows exponentially over a few characteristic Debye lengths. As the plasma becomes denser and the central channel longer, the sheath becomes compressed; if smaller than the mean free path for collisions, all electrons leaving the cathode will gain the maximum energy which can be supplied by the field as they accelerate into the plasma channel. Here the resistivity is now low and the electrons can freely flow to replace those leaving at the anode end. As the current flows, the resistance drops further and a high current may flow lowering the potential difference between anode and cathode. The external circuit now plays the decisive role as to whether the discharge stops (i.e. no charge left on plates) or continues (i.e. current can be supplied to sustain an arc).

Pseudospark

If the pressure is reduced, the mean free path of the gas increases and the possible kinetic energy of the electrons increases, meaning that a collision with a molecule becomes more likely to result in ionisation. This is the case until a point is reached where the gap separation becomes similar to that of the mean free path of the electrons. If the pressure is reduced further (or the gap reduced) more and more electrons will reach the anode without taking part in any interaction. A greater voltage is required to achieve breakdown. This behaviour is represented by Paschen's law (see Appendix). Figure 2.3 shows Paschen curves (plots of breakdown voltage between two electrodes with a separation, d, and a gas pressure, P) for plane parallel electrodes in air, nitrogen, and hydrogen. The area to the left of the dashed line is known as pseudospark, to the right spark discharges take place and around the minimum is the glow discharge region (where e.g. fluorescent lights work). The strange concept, which arises in the pseudospark range, is that longer distances discharge more easily, or with a lower voltage than short distances, and it is on this fundamental principle of ignition that the HCT lamp is based.

A fast discharge between a hollow-cathode and an anode at low pressure was named as "pseudospark" in 1979 by Christiansen and Schultheiss in their paper[18]: „Production of high current particle beams by low pressure spark discharges". In the late 1950's Christiansen worked on parallel plate avalanche counters for nuclear physics experiments. Near the Paschen minimum, anomalous sparks at the edge zones of these counters resulted in erratic failures and the discovery of the pseudospark.[19]

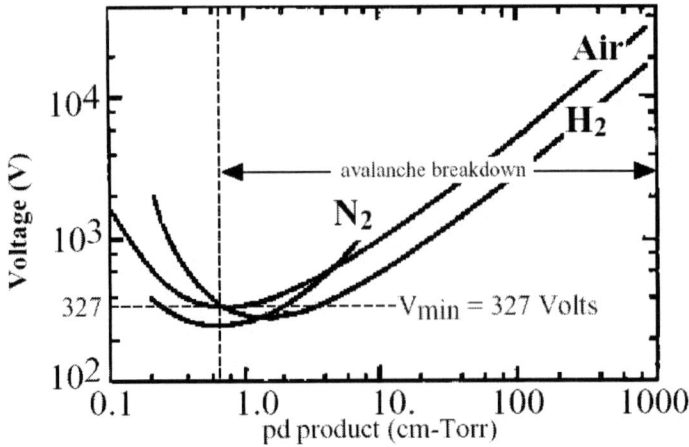

Figure 2.3: Paschen curves showing the required breakdown voltage, V, over a separation, d, and a pressure, P for air, nitrogen, and hydrogen.[17] **The Paschen minimum of 327 V for air is marked.**

2.2. Hollow Cathode Discharges

The system used in the HCT lamps is a discharge across two electrodes with a special geometry. This is shown in figure 2.4. A gap separation of two parallel electrodes, d_1, is on the order of a few mm and with a pressure of around 20 Pa the operation is on the steep negative slope side of the Paschen curve, the pseudo-spark region. The addition of two holes in the electrodes allows a discharge over a longer path, d_2. The hole in the anode is not absolutely necessary but does, in our case, allow the light to be extracted.

Assuming the holes are quite large (i.e. $\gg d_1$), then electrons starting on the parallel face of the cathode will mostly make it to the anode with no interaction, contributing only to a dark current, whereas electrons starting from, or near, the walls of hollow-cathode will interact a couple of times allowing a soft "glow discharge" to build up on the axis of the two holes. How quickly and how intense this takes place will depend on many other parameters such as the ratio of d_1 to d_2 or gas pressure and type.[53]

If now we consider the case where the hole is very small, it is fair to compare this to having no hole at all. In the hollow-cathode there will be no field and a discharge cannot take place. In the range between these two extremes (i.e. when the hole has a diameter comparable to d_1), a dynamic process takes place, which is known as a hollow-cathode effect.

This plasma ignition effect is of special interest when compared to other systems such as the plasma focus, the z-pinch or capillary discharges. The hollow-cathode effect allows a temporal build-up of low resistive plasma, thus allowing a pulsed current in the electrode system and the omission of a switch between electrodes and the capacitor banks. The processes and parameters involved require greater physical understanding but the concept allows sophisticated pulsed power technology for the energy supply to be omitted.

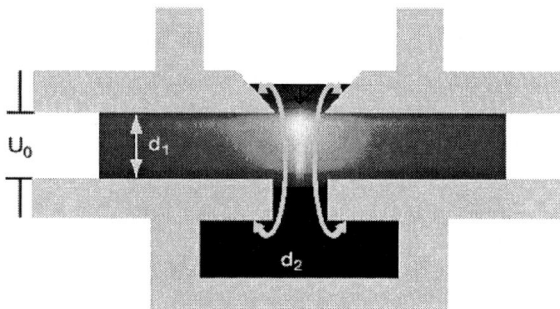

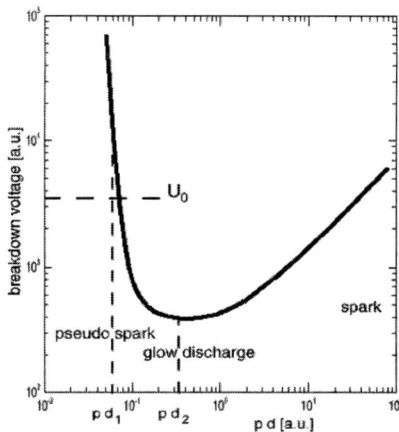

Figure 2.4: Above: The pseudospark geometry. Below: The corresponding Paschen curve. The gap d_1 is too small to discharge, whereas d_2 is much more favourable.

The parameters relevant to the ignition are the gas density, the main voltage and its rise time, the geometry of the lamp, the electrode material and also the initial conditions, such as the plasma density remaining from a previous shot.

Several works have already modelled the plasma ignition, which give a good qualitative description of the discharge phases.[20), 56), 57)] However these do not cover aspects such as operating in xenon or the addition of a trigger electrode, which makes comparison to the present problem difficult.

The discharge process can be considered in several phases.[20)] These phases are described in the following sections and are supported by the results of models made using Monte Carlo software with data more typical of the HCT lamp.[43)] As electrons achieve high energies due to large mean free paths, a usual fluid model is not valid for proper description of ignition phase. The model is used to solve the equation for the distribution functions, f_i, of electrons and different ion species:

$$\frac{\partial f_i}{\partial t} + \vec{v}\frac{\partial f_i}{\partial \vec{r}} + \frac{q_i}{m_i}\vec{E}_{mean}\frac{\partial f_i}{\partial \vec{v}} = \sum_k C_k f_k \qquad (2.5)$$

where C_k is a collision operator, which incorporates processes such as electron impact ionisation, elastic electron ion collisions and electron impact excitation into account. The particle motion is simply described by:

$$\frac{d\vec{v}}{dt} = \frac{q_i}{m_i}\vec{E}(\vec{r}) \qquad \frac{d\vec{r}}{dt} = \vec{v} \qquad (2.6)$$

where $\underline{E}$, is the self-consistent electrical field generated by the plasma itself and the voltage applied to the electrode system. Self consistently solving the Poisson equation,

$$\Delta\Phi = -\frac{e}{\varepsilon_0}(n_i - n_e) \text{ and } \vec{E} = -\vec{\nabla}\Phi, \qquad (2.7)$$

allows the electrical field to be calculated and the collisions are considered as random processes where a pseudo random number determines the outcome. For the simulation a moderate main voltage of 1·5 kV is used (a moderate voltage slows down the process, allowing more accurate calculation), a neutral xenon pressure of 26 Pa (experimentally found to be suitable for EUV production) and an initial electron and ion density in the system of 10^{12} m^{-3} to represent the products of a previous shot. Results are shown and discussed in the next three sections.

2.2.1. Townsend Phase

When the voltage is first applied, the electric field is barely affected by the presence of the small hole in the cathode. Electrons and ions will drift across the gap as with the parallel plate geometry forming only a dark current. In the region of the hole, electrons and ions also begin to form currents, yet the ions here do not meet the cathode and release new electrons but rather, they drift on into the cathode chamber. Naturally present free electrons (created by cosmic rays or radiation background) in the hollow-cathode chamber may also leave via the hole and be attracted towards the anode. The hole could be described as a one-way trap; electrons can't go back in, nor do ions come out.

Initially after applying the potential, the space charge of the ions and electrons is small and has little effect over the dominant electric field and is therefore still not affected by the presence of the hole; the hole's diameter is small compared to area the electrode face.[21] Electrons are accelerated in the field towards the anode without collision, forming only a small electron beam on the axis. Most of the initial ionisation processes take place in the hollow anode near the borehole and along the axis within the gap as shown in the 5 μs Monte Carlo diagram. Most of the electrons are rapidly extracted, leading to an excess of ions, the exception to this being the main gap region, where all initial electrons *and* ions have already been extracted.

As more and more electrons leave the hollow-cathode, and ions enter, the field will begin to distort due to the positive charge building up. The few electrons that do take part in collisions will do so close to the anode (the probability of collision is greater after they have crossed the gap). After such a collision, the electrons will be rapidly lost to the anode and the ions will slowly start drift to the cathode side. This positive charge pushes the field away from the anode, increases the field strength and hence the multiplication. This phase so far is similar to the avalanche process described previously, where self-breakdown can be used to initiate the process; i.e. cosmic rays, radiation background, secondary electrons from ion collisions or the field emission from the cathode. At this stage the resulting maximum ion density from the simulation is $8\cdot4.10^{14}$ m^{-3} whereas the maximum electron density is $0\cdot14.10^{14}$ m^{-3}. To successfully continue to the next phase between 10^9 and 10^{11} electrons are required to initiate the process.[20] Alternatively triggering mechanisms may be applied to supply these and to improve the temporal stability.

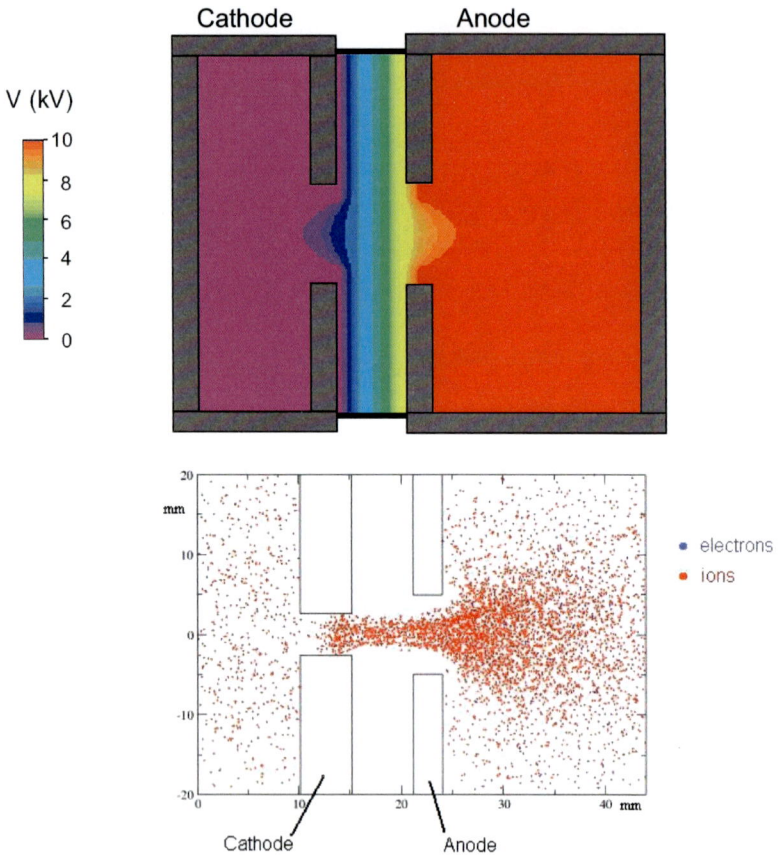

Figure 2.5: Above: Initial stage of the discharge. The hole is too small to have an effect on the field. The colours represent qualitatively different equipotential ranges.

Below: Spatial distribution of ion and electron densities after 5 μs: result from the Monte Carlo simulation.[43]

2.2.2. Hollow-Cathode Phase

As the field is pushed towards the cathode, becoming ever more intense, "runaway" electrons are more effectively accelerated and emitted as a beam towards the anode, some of which contribute through their collisions towards the positive charge building up at the anode side. Together with the effect of the space charge inside the hollow-cathode, the electric field enters through the hole into the hollow-cathode, causing a peak (to about 100x) in the electron multiplication.[20] The maximum electron and ion densities now increase to $n_{e_{max}} = 7 \cdot 6.10^{14}$ m^{-3} and $n_{i_{max}} = 4 \cdot 2.10^{15}$ m^{-3} respectively for the 15 μs case, see figure 2.6.

A more intense beam is now emitted as electrons are accelerated towards a focus in the anode. This is due to the converging electric field, which progresses further into the hollow-cathode; this beam from the hollow-cathode is capable of reaching energies of several keV, in the energy range of the applied voltage and have been known to reach 10 to 20 % of the discharge current,[20] depending on the geometry. For this reason lead glass is used for the windows of the vacuum system as, x-rays could be produced by the beams on the glass. Hollow-cathode arrangements have been used as sources of charged particle beams.[18]

Electrons now emitted from inside the hollow-cathode accelerate in the field which is present and the ionisation processes shift more and more towards the hollow cathode ionising more of the neutral gas forming more intense plasma on the axis of the holes. As this plasma penetrates into the hollow-cathode, the potential on the axis becomes very close to the anode potential and the Monte Carlo result (see figure 2.7) shows the formation of the "virtual anode" within the hollow cathode after 21·2 μs in the lower figure. The simulation now shows maximum electron and ion densities nearly identical at around 8.10^{17} m^{-3}. At this point, the simulation terminated, as the description of further development including the hollow-cathode plasma and onset of the pinching phase requires some numerical problems to be solved.

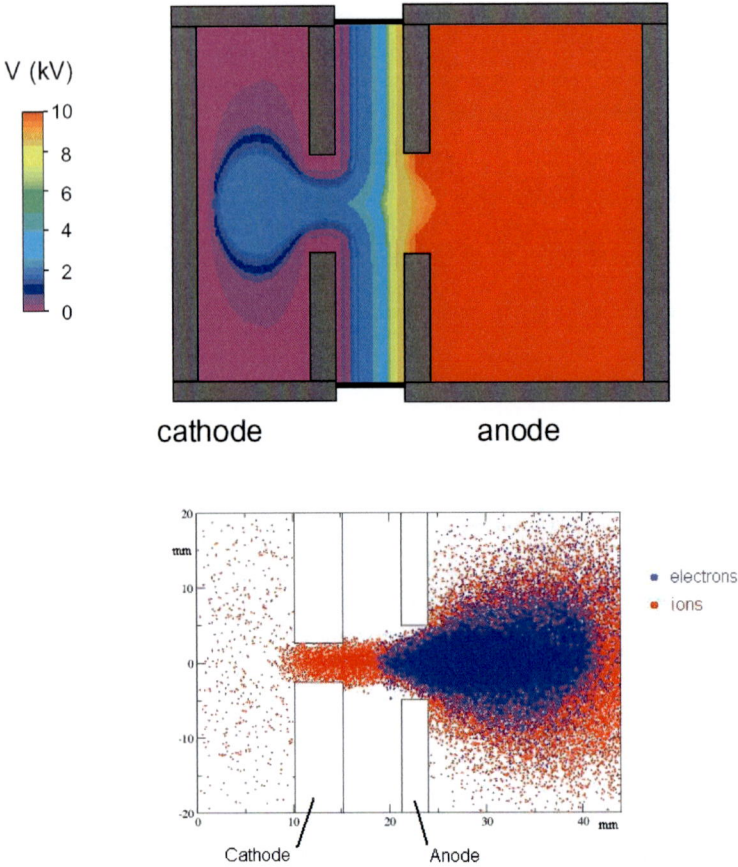

V (kV)

cathode   anode

electrons
ions

Cathode   Anode

Figure 2.6: Above: Qualitative figure of the formation of plasma allowing the field to distort and enters the hollow-cathode due to the positive space charge.

Below: Spatial distribution of ion and electron densities after 15 µs: result from the Monte Carlo simulation.[43)]

The shape of the field at this point allows a peculiar effect to take place: That is, that electrons can accelerate from one side and then be reflected at the other, so to say, like a "pendulum" from side to side inside the hollow-cathode. This increases their journey well over the mean free path for collision and likewise their ability to ionise the gas further.

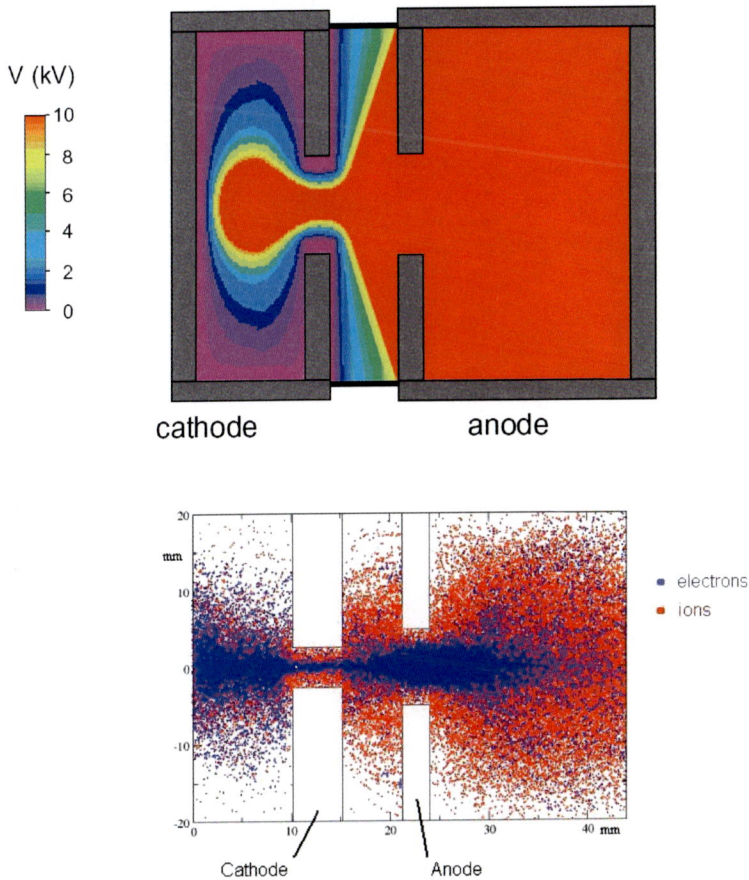

Figure 2.7: Above: Qualitative figure of the "virtual anode" fully inside the hollow-cathode.

Below: Spatial distribution of ion and electron densities after 21 µs: result from the Monte Carlo simulation.[43]

Avalanches can now form and the multiplication factor therefore rises sharply, as does the discharge current. Each of the electrons can traverse the cathode cavity several times before being extracted through the cathode hole. Most of the energy gained by the electrons is now given up heating the plasma, rather than to the surface of the anode as it was previously.

2.2.3. Bottleneck Phase

If the pressure is high enough, the hollow-cathode phase will end and this coincides with the termination of the previously intensifying electron beam. If the pressure is low, the system can remain in this dense-glow phase (until the capacitor banks have been emptied), as is the case for electron beam sources.[22] The condition found to define this pressure threshold is referred to as the "critical pressure" or "critical particle density", n_{cr}.[23), 24), 25)]

$$n_{cr} = \frac{1}{\sigma_i d} \sqrt{\frac{3m_e}{M_i}} \qquad (2.8)$$

where m_e and M_i are the mass of the electrons that of the ions respectively, d is the gap separation and σ_i the ionisation cross-section for the atoms by the electrons. Essentially what this states is that the number of ionisation events, created by one electron at the anode, $n_{cr}s_id$, is proportional to the inverse square root of the ionic mass (in other words heavy gases require higher pressures to form an arc). Assuming this condition can be fulfilled, the borehole becomes blocked by plasma, which electrically shields the gap from the hollow-cathode cavity. The dense plasma in the cathode and hole causes ion sheaths to form around the walls.

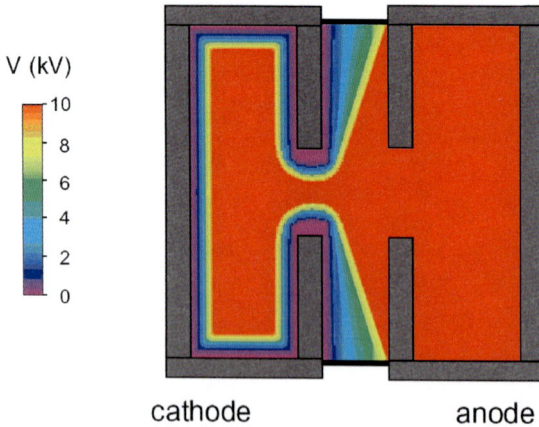

Figure 2.8: Qualitative figure of the plasma expansion in the hollow-cathode.

Current can now no longer leave via the hollow-cathode chamber but rather now from the surface of the hole. A self-sputtering processes[26] of ions bombarding the cathode removes some cathode material, even in light gases such as hydrogen.[26] This allows the multiplication to continue increasing sharply. Neutral metal sputtered away from the surface becomes ionised, as the mean free path for ionising these fast sputtered metal atoms is less than gap. These ions are then accelerated back to the cathode[27] and the process can repeat. A pendulum effect, now with the ions in the hole allows the plasma density at the cathode side increase exponentially[26] and electrical breakdown can enter the high current phase.

2.3.4. High-Current Phase and Pinch Formation

Highly conducting plasma now fills the system and conducting channels can form. The plasma, from which the current forms, is subject to its own magnetic field, which circles around the current axis, in particular in the region of the hole, as this is where the multiplication was the strongest. Charged particles moving in a magnetic field are subject to the Lorentz force - perpendicular to both the field and the current, as shown in figure 2.9.

Figure 2.9:   The Lorentz force, F (F = qv x B), acting on a positive ion or a negative electron (or ion) with a velocity, v, in the presence of a magnetic field, B. With a field going into the page a charge, q, is accelerated downwards if it is moving to the left. Likewise for a negative charge moving in the opposite direction.

The field around the discharge supplies energy which has the effect of confining the plasma to the axis of the current, i.e., it "pinches". The pressure of the magnetic field created by the current is:

$$P_{mag} = B^2/2\mu_o \qquad (2.9)$$

where B is the magnetic field and μ_o the permeability of free space. Assuming that the pinch may be considered to be in equilibrium, the confining magnetic pressure is balanced by that of the plasma

$$P_{plasma} = n_e\, kT_e + n_i\, kT_i \qquad (2.10)$$

proportional to the number of free electrons (or ions) per unit volume, n_e (n_i), Boltzmann's constant, k, and the temperature, T. So, if the theory for a Bennett Pinch[28] is assumed, and the magnetic field is only generated by the current in the plasma, the two equations can be solved for pressure

$$B^2/2\mu_o = n_e\, kT_e + n_i\, kT_i \qquad (2.11)$$

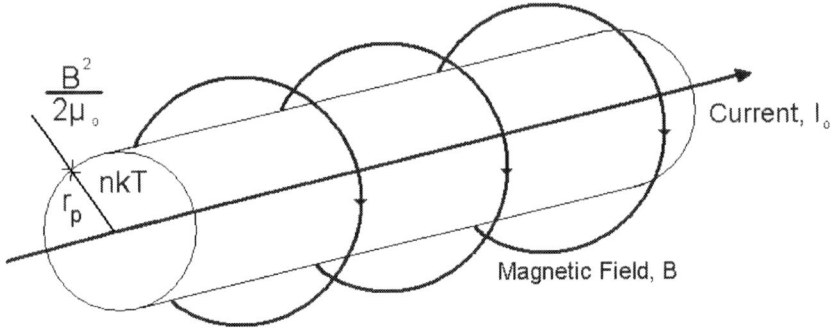

Figure 2.10: The pinch plasma is considered as a cylinder of radius r_p in equilibrium on the axis of the current between the two bore holes.

Using Ampere's law, the magnetic field around the pinch at a radius r_p can be found as

$$B = \mu_o\, I_o\, /\, 2\pi r_p \qquad (2.12)$$

Together with equation (2.11), if the required electron density and electron temperature of the plasma are known (or can be estimated), a value for the discharge current can be achieved

$$I_o^2 = \frac{8\,k\,T_e\,\pi^2\,r_p^2\,n_e}{\mu_o} \qquad (2.13)$$

where it is assumed that T_i is small compared to T_e and the second term on the r.h.s. of (2.10) is neglected. An estimate for the electron density can be made from the neutral gas properties. The number of available xenon atoms at room temperature and 10 Pa pressure is found to be

$$n_{Xe} = \frac{P}{kT} = \frac{10\,\mathrm{Pa}}{1\cdot38.10^{-23}\,\mathrm{J.K^{-1}}.293\,\mathrm{K}} = 2\cdot5.10^{21}\,\mathrm{m}^{-3}. \qquad (2.14)$$

Considering the compression of the plasma to be from the neutral gas within the bore hole radius of 5 mm down to the observed radius of the pinch of 0·5 mm, with a constant length equal to the gap separation, the electron density can be estimated to be around

$$n_e = 12.n_{Xe}.\left(\frac{r_o}{r_p}\right)^2 = 12.2\cdot5.10^{21}\,\mathrm{m}^{-3}.\left(\frac{5.10^{-3}\,\mathrm{mm}}{0\cdot5.10^{-3}\,\mathrm{mm}}\right)^2 = 3.10^{24}\,\mathrm{m}^3 \qquad (2.15)$$

For a plasma to emit in the region of 13·5 nm, using Wien's Law, its temperature can be found from:

$$T = \frac{\sigma_W}{\lambda_{max}} = \frac{2\cdot898.10^{-3}\,\mathrm{mK}}{13\cdot5\,\mathrm{nm}} = 2\cdot2.10^5\,\mathrm{K}. \qquad (2.16)$$

Therefore, a rather hot plasma of around $2\cdot2.10^5$ K is required for 13·5 nm radiation to be emitted. Using the information found in equations (2.15) and (2.16); equation (2.13) can be solved for the required discharge current,

$$I_o^2 = \frac{8.k.T_e.\pi^2.r_p^2.n_e}{\mu_o} = \left(\frac{8.1\cdot38.10^{-23}.2\cdot2.10^5\pi^2\left(0\cdot5.10^{-3}\right)^2.10^{24}}{4\pi.10^{-7}}\right)A^2 \approx 5.10^7\,A^2 \qquad (2.17)$$

$$I = 7\,\mathrm{kA}$$

This would correspond to a power of 0·3 MW, given a pinch resistivity of 6·6 mΩ (see equations 6.4 – 6.6). Obviously, this kind of power in the required volume of a few cubic millimetres puts CW operation of the lamp out of the question and a pulsed system must be used.

As the high current flows, and the plasma pinches, it is heated to higher temperatures, dependent on the current that flows. As the plasma cools and decays, EUV radiation can be produced. Figure 2.11 shows a typical spectrum for xenon and the 13·5 nm peak.

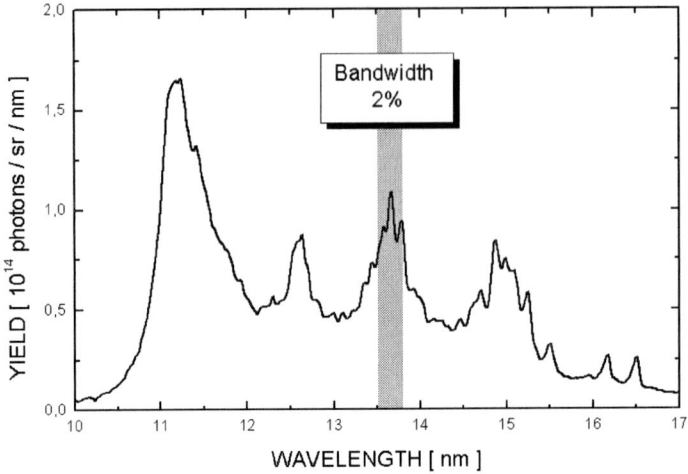

Figure 2.11: Spectrum of xenon in the EUV region showing the 13·5 nm peak and the 2% bandwidth which can be collected by the optics.

3. Triggering the HC Discharge

Many possibilities exist for triggering a pseudospark discharge.[13), 29), 30)] Triggering is intended stabilise the ignition conditions, bringing pulse stability in both time and amplitude, avoiding missed shots or early breakdown. Over the untriggered lamp, it should also make the repetition frequency independent from the charging current. The triggering device is added to the pseudospark geometry and can operate in two ways, either actively, causing a discharge to take place in a system not normally capable of discharging, or passively, preventing the discharge from taking place in a system which is willing to discharge.

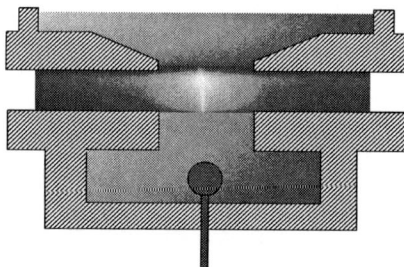

Figure 3.1: **A trigger device (shown schematically) can be used to discharge or block the system. Several possibilities for what this device can be are discussed below.**

This work concentrates mainly on the case where the trigger mechanism is a third electrode added to the hollow-cathode, although other active triggering methods are also discussed. The optimisation of the form of this electrode and exactly how it will function are then the main aim of the work. The expected effect of the trigger electrode in the pseudospark-glow-discharge range is marked in figure 3.2. When a positive potential difference is applied using a low-power switch between the trigger and the cathode, it is expected that a greater voltage be required to achieve breakdown. Similarly, when a negative voltage is applied, this is expected to supply electrons, assisting the discharge.

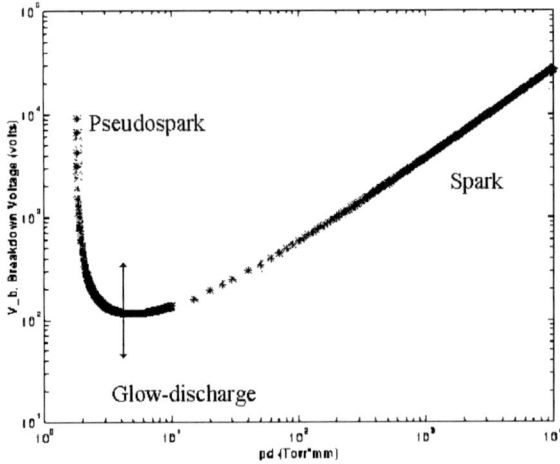

Figure 3.2: Applying a constant voltage to the trigger electrode should shift the Paschen curve up if positive or down if negative in the pseudospark and glow-discharge regions.

3.1. A Literature Review

As all previous knowledge and data applies to pseudospark switching devices rather than lamps, the following sections will show what trigger methods are available and compare their suitability for use in a lamp. In some cases, similar systems are built and tested in the lamps to identify similarities and problem areas. As pseudospark switches are used for pulsed power-applications, it is already identified that the energy in a switch system is dissipated in an external load rather than in the electrode system itself; as is the case for a lamp. Also further boundary conditions come in to play such as which fill-gas is used and the pressure region where EUV can be produced.

3.1.1. Glow-Discharge Triggering

With active triggering it should be possible to vary parameters such as pressure and discharge voltage over a very wide range, as the required amount of ionisation can be created to initiate the discharge whatever the conditions. Of particular interest is still the left-hand side of the Paschen curve, but now at much lower pressures than defined by the

Paschen curve. By operating the lamp to the left of the Paschen curve a higher maximum frequency is expected, as there is no longer the risk of an early breakdown in the gap as the frequency is increased.

An active trigger was used to operate a pseudospark switch.[31] This is shown in figure 3.3. A forth electrode, B, is added to the system to allow a glow discharge to burn continuously in the rear chamber (AB) powered by an auxiliary supply. This glow forms a plasma in the rear chamber. The main gap is formed by electrodes C and D with the main cathode cavity, C, grounded and this represents an anode with respect to the glow electrode, B.

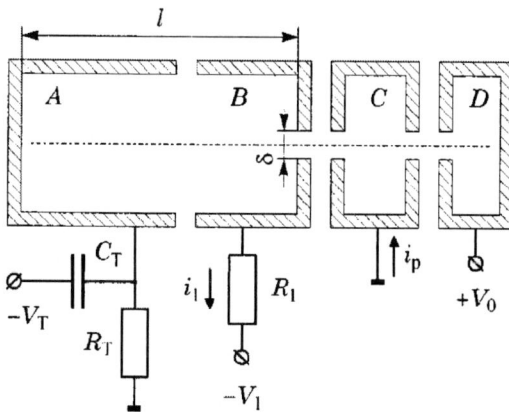

Figure 3.3:[31] Schematic of the electrode system of the switch and the electric circuit. The main gap spacing, d and the thickness of the electrodes is 3 mm and the inner diameter of the electrodes A and B is 3 cm. Up to 3 kV was used to ignite a glow of 1 mA.

Some fraction of the glow current (<10 %) flows to the cathode cavity and the polarity of this current can define whether the hold-off voltage of the main gap is increased or decreased. The switch is triggered when a negative trigger pulse, V_T, is applied to the trigger electrode, A, relative to the grounded electrode C. The voltage must be high enough to rearrange the discharge in the trigger chamber so that the triggering discharge between A and C generates enough plasma in the cavity as a source of intense electrons for the main gap discharge[32]. It should be noted that the long distance, l, is required to keep the glow discharge burning.[13]

The principle of intercepting the auxiliary glow discharge current in the trigger electrode system and diverting it to the main cathode cavity allows the switch to be triggered at a low voltage and current level. Two modes of discharge are described representing the cases where the discharge current is above or below 2 mA. If the glow current is too low, the trigger discharge cannot close properly to the electrode C through the borehole and the system becomes unstable. A trigger current of about 0·25 A was used to operate up to several kilohertz with the delay time to triggering not exceeding 200 ns.

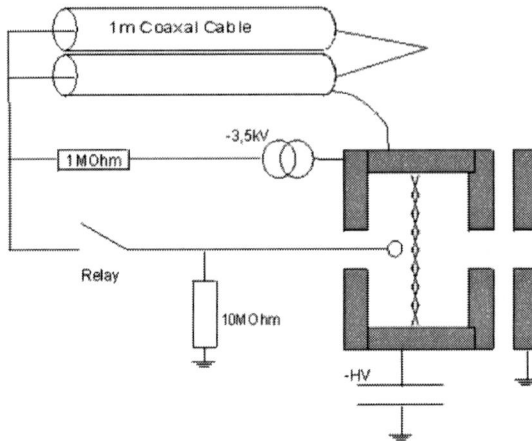

Figure 3.4: Circuit of the glow-discharge trigger set-up. The metre of cable serves as a capacitor to protect the power supply.

Figure 3.4 shows the circuit of our experimental active trigger set-up. The lamp is the same as non-triggered version, except an electrode is placed centrally in the back wall behind a metallic mesh. The external circuit has been modified to give large negative pulses from a power supply connected relative to the cathode. When the main voltage is applied, the electrode remains at ground potential until a glow discharge starts to burn in the hollow cathode behind the mesh. The relay then closes and the negative high voltage pulse is fed to the electrode. This large voltage spike should accelerate free electrons through the mesh, much the same as an electron gun in a television set. Experiments were carried out on a low power machine (~10 Hz limited by the relay); nothing else was changed to lamp's electrode geometry.

With the glow discharge present, the required breakdown voltage was clearly lower than without, however the Paschen curve was difficult to measure, as the lamp itself enters an untriggered glow discharge mode. The measured curve without the glow present is shown in figure 3.5 that the result lies safe from small-gap discharges on the steep left-hand pseudospark branch of the Paschen curve, a 2 Pa change in pressure corresponding to about double the required breakdown voltage. This reflects the untriggered lamp's extreme sensitivity. The dashed line represents the approximate position of the Paschen curve with the glow present.

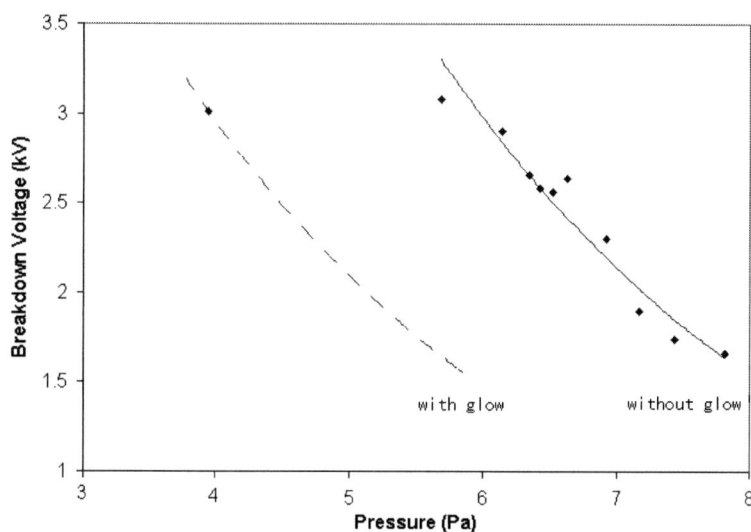

Figure 3.5: Paschen curves of the lamp with the glow-discharge trigger present and without.

As the problems in measuring a Paschen curve show, the trigger chamber must be less coupled to the hollow-cathode. In general, the glow-discharge trigger disturbs the main gap, lowering the required pressure for breakdown too far. Its presence was only intended as a source of charge carriers for the trigger to inject to the hollow-cathode. A modified version of the active trigger is covered in the following section.

3.1.2. Discharge Triggering with Blocking Electrodes

At low frequency glow-discharge triggering functions quite successfully.[13], [29], [30] Increasing the size of the trigger chamber allows a faster discharge to take place; the products of which then go on to ignite the main discharge. The addition of passive blocking electrodes also allows a system to operate above its usual self-breakdown pressure. A combination of this with an active trigger should widen the available pressure range in both directions and gain access to high frequencies and high lifetimes.[33], [34] A pseudospark switch triggered by a pulsed low-current gas discharge was developed in (33) capable of switching a high voltage of up to 20 kV and currents of 10 kA. The switch worked at high repetition rates of up to 100 kHz and fast rates of current rise times of up to 5.10^{11} A.s^{-1}.

The jitter was reported to be much smaller than 1 ns and the pseudo-spark switch was successfully tested with two external loads running in place of a hydrogen thyratron and a high-pressure spark gap switch.

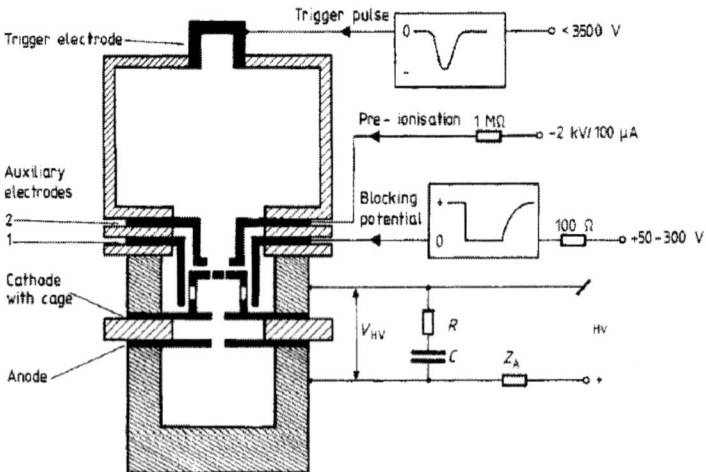

Figure 3.6: Schematic of a combined active trigger with blocking electrodes for a pseudospark switch. The main capacitor bank, C, has a value between 2 - 6 nF and discharges through the load, R, in this case a laser.[33]

The pseudo-spark switch and its electrical circuit parameters are shown schematically in figure 3.6. The construction made of brass and glass had hydrogen, nitrogen or argon as the working gases. The main anode-cathode separation is about 4 mm, and the other switch dimensions to the same scale, make this device quite small compared to other systems. The discharge region is separated from the trigger module by a cylindrical cage forming the hollow cathode. This cage screens the trigger section from the main discharge, providing a practically unlimited lifetime of the trigger module.[33] A positive voltage of 50-300 V is applied to the blocking electrode.

A construction similar that in (33) was built by modifying the electronics of the previous glow discharge trigger and replacing the relays with a high-voltage semiconductor switch (Behlke) to allow scaling-up to higher frequencies. For simplification of construction, the same electrode is used for the glow and the trigger discharge. The hollow-cathode back wall has four holes, each 3 mm, instead of a mesh, to de-couple the trigger chamber from the hollow-cathode during charging. Smaller holes reduce the trigger effect too much for stable operation of the lamp.

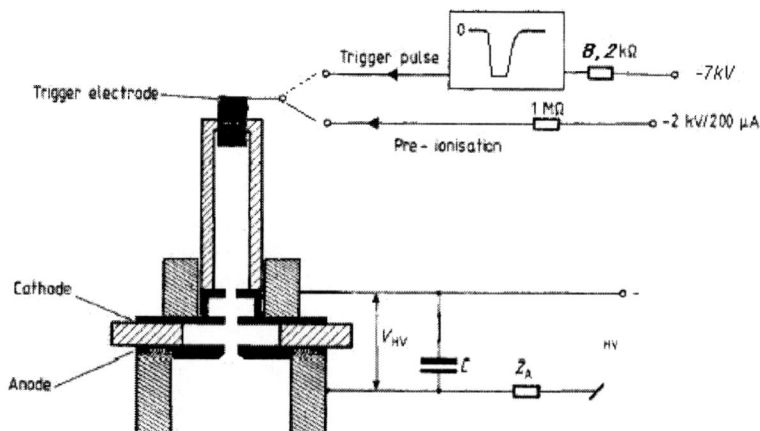

Figure 3.7:   Schematic of the high-power discharge trigger tested for use as a lamp. A 60 mm trigger chamber is coupled only by four small holes to the hollow-cathode, allowing plasma decay in the hollow-cathode even if a glow is kept burning at the trigger.

The main gap is now much less disturbed by the glow (if used) until the trigger discharge takes place. The operation of the lamp can be seen in figure 3.8. The trigger potential of -8 kV is applied in ~800 ns and after a 2·5 μs delay the lamp and trigger discharge simultaneously. However, the inside of the ceramic tube was heavily blackened (metallised), this took place over only a few minutes of operation at 1 kHz (~10^5 shots), a problem not noticed in the Hz range before. In the case of a metallised ceramic, the trigger sometimes surface discharges or breaks down late directly to anode potential, running the lamp with a large jitter and very unstable operation.

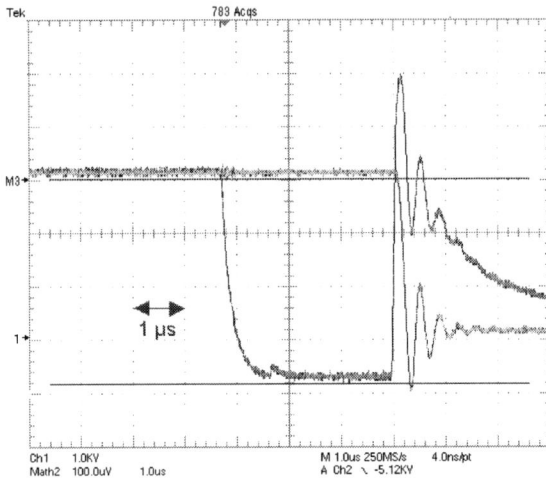

Figure 3.8: **Voltage profile of the discharge trigger. Ch1. - the main gap voltage (-3 kV, inverted signal). M3. - calculated voltage on the trigger electrode relative to the cathode potential.**

By using clean ceramic and running the lamp only for short periods, measurements could be made to determine the best conditions for up-scaling (pressure, gas flow, geometry, voltages). One large influencing factor found using this trigger is the capacitance; not of the external electronics, but rather the trigger electrode in the lamp. The delay, between applying the trigger and the discharge, depends not only on the trigger voltage, but also on how quickly the trigger reaches its potential (-7 kV). If the potential is applied slower than 10^9 V.s^{-1}, the ignition time tends to infinity.

Triggers with large areas increase the delay time and therefore reduce the stability, although their large surface area may be advantages for ignition and current dispersal over the surface. By reducing the inductance and capacitance of the trigger (~2 pF) the lamp can be operated with reduced delay and the jitter. The trigger should be applied in less than 1 μs to give the most stable operation. As far as lifetime is concerned, it can be concluded that large ceramic surfaces must be avoided in the lamp due to metallisation.

3.1.3. Dielectric trigger

To avoid changing the geometry of the system and the need for long distances to ignite a trigger glow, a dielectric field emission trigger may be used.[13] This is achieved by placing an electrode inside a thin insulator with a high dielectric constant behind a copper mesh which makes up one wall of the hollow-electrode. Such a system was used to simultaneously trigger up to 12 hollow-cathode discharges with delay times of less than 150 ns.[13] The trigger effectively forms a capacitor with a hollow-cathode allowing only a fast trigger pulse of (~10 kV and ~1·2 mA) to take place and no DC current.

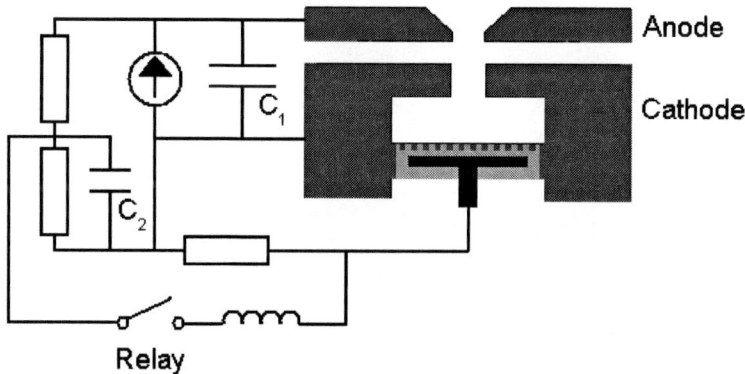

Figure 3.9: **Schematic of the dielectric trigger set-up. The power supply charges C_1 with a constant current. The two resistors on the left form a potential divider and set the voltage on C_2 ready for the negative pulse, while the third resistor allows the trigger to recover after the pulse.**

The switch required neither a heated cathode or a glow discharge. The breakdown initiation is direct, in that the initial number of electrons is greater than 10^{13} cm^{-2}, so no additional multiplication is necessary to commute the switch resistance from near infinite to a low value in a short time.[35] The electrons are ejected from the dielectric material in the low pressure range 1 - 50 Pa[13], which cause ionisation and the onset of the hollow-cathode effect. Lifetimes of up to 10^7 shots are also reported for such a trigger mechanism[33], [13]. Using H$_2$ as the working gas and currents of up to 400 kA could be switched.[36), 33]

Figure 3.9 shows a similar construction built for comparison and to test this method's suitability as a trigger. The capacitor, C1, was charged to at least 1·5 kV and then switched via a coil to the dielectric trigger. A current of around 10 A flows instantaneously to the trigger electrode initiating the discharge of the lamp. Using the computer simulation software (Pspice) an equivalence circuit was used to predict the current and voltage waveforms for the trigger. These are shown in figure 3.10.

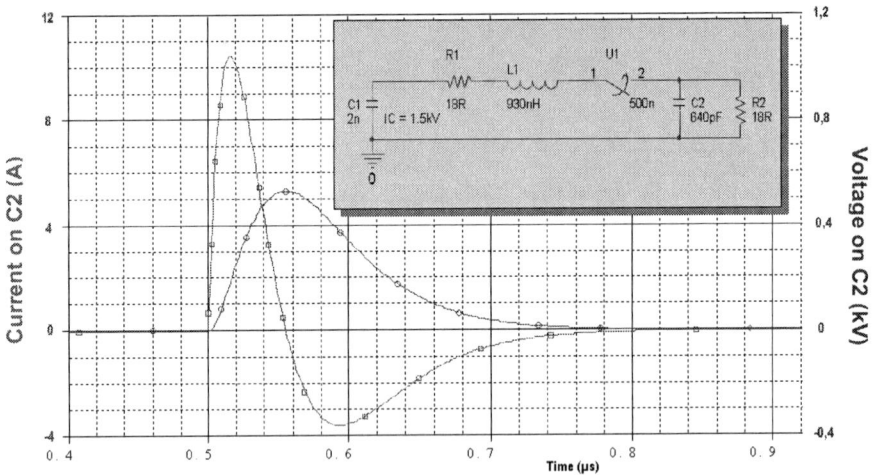

Figure 3.10: **Equivalence circuit (inset) entered into "Pspice" to simulate the dielectric trigger. Plotted are the calculated voltage waveform (circles) and current flowing to the trigger capacitance (squares).**

In comparison, the trigger improved performance over that made with a self-breakdown discharge. Constant trigger voltage was used and the pressure could be varied over a range from 2 to 12 Pa. However, high-pressure combined with high-repetition rate could not be achieved. A maximum in the EUV signal was found towards lower pressure, which is shown in figure 3.11.

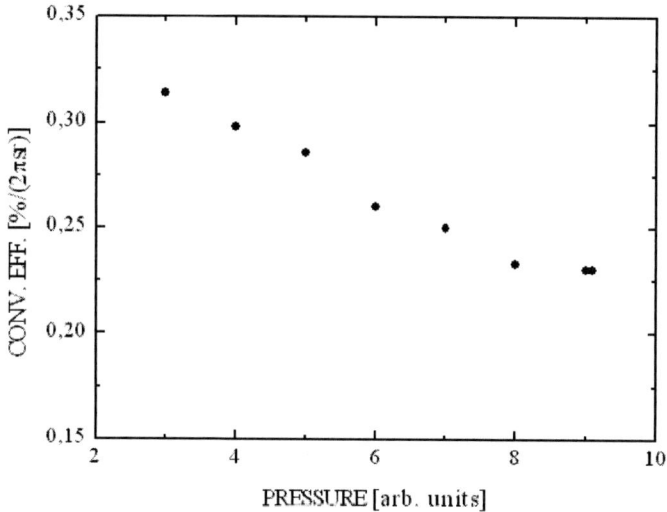

Figure 3.11: Variation of the EUV conversion efficiency as a function of gas pressure using the dielectric trigger. The maximum is found at lower pressure with less energy.

3.1.4. UV Flash-Lamp Trigger

If a pulse of UV light is given inside the hollow-cathode, photoelectrons will be emitted from the electrode surface.[15] These electrons are then available as seed electrons for a hollow-cathode discharge. Such a system was used to switch currents of over 17 kA at repetition rates of 100 Hz in a glow mode with a breakdown voltage up to 37 kV.[37]

The number of electrons released can be found from the energy of the UV light incident on the cathode. Assuming the light monochromatic with a frequency ν, then n_p photons will be present on the electrode surface

$$n_p = \frac{E_{uv}}{h\nu} \tag{3.1}$$

where h is Planck's constant and E_{uv} the light energy incident on the surface. From the surface n_e electrons will be released proportional to the electron emission coefficient γ_f of the electrode material

$$n_e = \frac{E_{uv}\gamma_f \lambda}{hc} \qquad (3.2)$$

where λ/c has been substituted for v. Taking a γ value of 0·1 and a flash lamp of 15 mJ with a wavelength of 300 nm, the number of seed electrons for the discharge is found to be

$$n_e = \frac{15.10^{-3}.0 \cdot 1.300.10^{-9}}{6 \cdot 6.10^{-34}.3.10^8} = 2 \cdot 3.10^{15} \qquad (3.3)$$

In a hollow cathode of a few cm^3, this corresponds to an electron density well in excess of that required for ignition.[36] In addition, the gas present may be ionised directly by the UV light forming a space charge. These ions will also be accelerated, colliding with gas atoms, contributing further to the ionisation process.

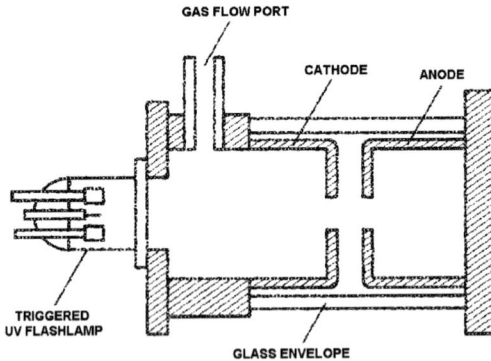

Figure 3.12 **Schematic of a flash-lamp-triggered switch.[37] Also known as a BLT (Back Lit Thyratron), photoelectrons are emitted from the inner cathode surface. The gap and holes are both about 2 mm with He or H$_2$ as the fill gas at ~10 Pa.**

3.2. Further Triggering Concepts

3.2.1. The Passive Triggering Concept

So-called "blocking" electrodes, held at a positive potentials, have often been used to prevent pseudospark discharges from taking place.[38] The role of this non-switched electrode was to extend the range of operating pressure for a pseudospark switch rather than to trigger the discharge. The reason for not using only a blocking electrode on its own probably stems from the desire to have a switch which operates over a wide range of voltages, and therefore a wide range of operating pressures. Blocking electrodes also have limited operation and can only be used up to a few hundred volts, as the electrode must remain well under its breakdown potential if it is to function effectively. Whilst this limitation is undesirable for a switch, it should not be a problem for a lamp, running at a fixed pressure and voltage. The concept of a lamp, triggered only by the removal of a blocking potential, is presented here.

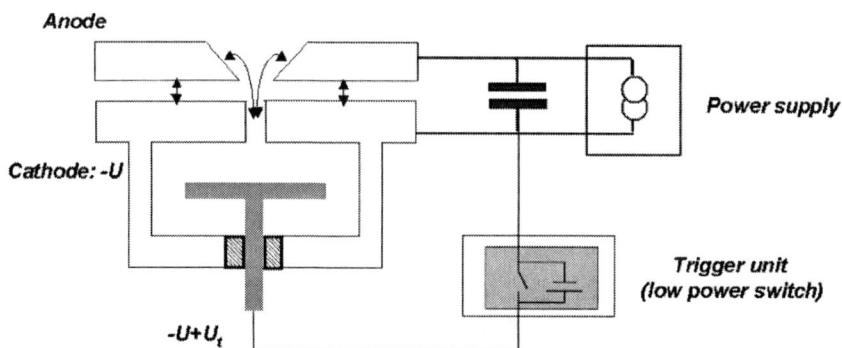

Anode

Cathode: -U

Power supply

**Trigger unit
(low power switch)**

$-U+U_t$

Figure 3.13: Schematic of the passive triggered system.[39] Breakdown occurs when the trigger potential is removed. When the positive voltage is applied to the electrode, a voltage higher than on the main capacitors is required to achieve breakdown.

A schematic of a lamp triggered by only a blocking electrode is shown in figure 3.13. The blocking electrode will be switched on with a few hundred volts and the lamp's main capacitors then charged to a voltage above the static breakdown potential. Due to the collection of the charge carriers, or other influence, the hollow-cathode effect is prevented from taking place. The lamp then fires immediately when the blocking voltage is switched off.

The expected parameter range of a passive triggered system is shown in figure 3.14. For a given voltage, V_{op}, on the lamps main capacitors, a pressure window is expected to be available, which stretches between the two curves. A pressure greater than the upper curve will lead to self-breakdown, and less than the left of the lower curve, will give no discharge at all. Similarly if a fixed pressure is chosen, P_{op}, then a voltage window is available where the lamp will function. It should be noted, that although the lamp will run anywhere between the two curves, the delay time between removing the trigger and the lamp discharging, Δt, will vary as a function of the voltage and pressure. The size of Δt and the temporal jitter is of vital importance when scaling up to high frequencies for lithography.

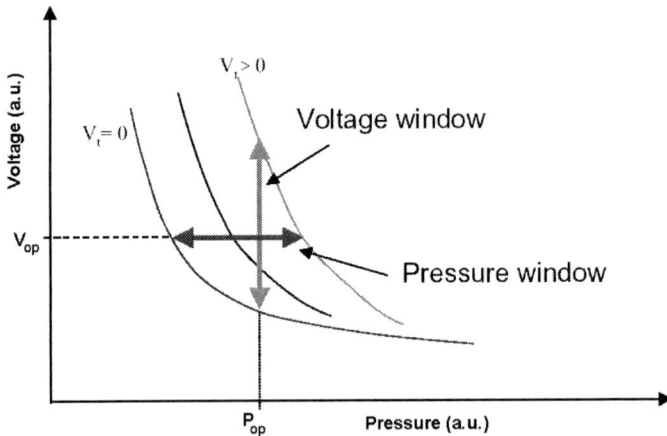

Figure 3.14: The expected available parameter range for the passive trigger. The lower curve represents the static breakdown (Paschen) curve, the upper, required potential to achieve breakdown with the trigger present and the central line, the suggested safe operation range for the lamp.

3.2.2. Microwave Triggering

Xenon plasma can be generated by the high-frequency electromagnetic fields from a magnetron.[40] Microwave radiation discharges have been used in the past for the excitation of noble gas lasers.[5] To use this principle to trigger the lamp, the hollow-cathode would have to form a resonator, designed, for instance, to provide a field maximum along the z-axis behind the cathode hole. If a standard 2·45 GHz transmitter is to be used (for economic reasons) the radius of the hollow-cathode would be defined by the frequency as

$$R = \frac{c}{2\pi} \frac{2 \cdot 405}{v} = \frac{3.10^8}{2\pi} \frac{2 \cdot 405}{2 \cdot 45.10^9} = 0 \cdot 047 \, \text{m} \qquad (3.4)$$

where c is the speed of light. This radius is somewhat larger than the hollow-cathode used until now, so a higher frequency may be necessary, as a large hollow-cathode limits the frequency of the lamp. Reversing the calculation, for the standard hollow-cathode of 10 mm radius, the frequency needed is found to be

$$v = \frac{3 \cdot 10^8}{2\pi} \frac{2 \cdot 405}{10.10^{-3}} = 11 \cdot 5 \, \text{GHz} \qquad (3.5)$$

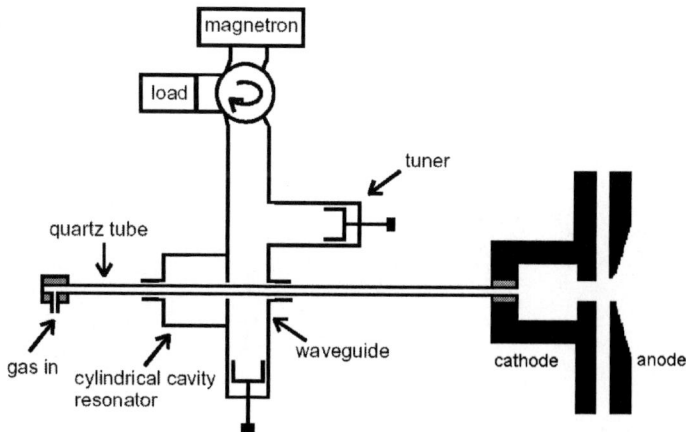

Figure 3.15: **A possible set-up of a microwave trigger. Plasma is generated in the capillary by microwave excitation which travels as a wave to the hollow-cathode. Here the small plasma jet triggers the hollow-cathode discharge.**

A possible simpler method would be to use a capillary,[5] passing through a μW-cavity resonator. A travelling wave would then be produced which could cover a large distance through an insulating tube to the hollow-cathode, where it emerges as a jet of plasma. This method would eliminate the need for isolation, as the hollow-cathode is at high potential, and limitations on the hollow-cathode geometry, as with electrical triggers.

3.2.3. Laser Trigger

Laser produced plasmas were introduced in section 1.4.2 as a possible source of EUV radiation. However, this method was considered to produce too much debris and the target material is used up. This could be used as an advantage though, for a possible combined triggering mechanism for the lamp producing EUV indirectly. A possible set-up could comprise of the standard hollow-cathode arrangement with a low-power laser aimed inside the cathode volume. When the laser is fired, a small amount of the electrode material forms a plasma which initiates the discharge of the lamp. The laser power could be kept as low as possible to minimise the amount of debris produced. If only a very low-power laser is applied and used to form a plasma in the hollow-cathode volume, then this plasma should have the effect of triggering the pseudospark discharge without sputtering away the target material too quickly.

This method makes the construction of the electrode system simpler and easy to cool, yet it is expensive and may be difficult to install in a commercial lamp, as the laser must sit in the EUV optical path. It does, however, open up many new possibilities, such as having no working gas at all; the cathode material could form all the material required for the pinch.

3.3. Discussion

The methods of active triggering presented in section 3.1 allow operation over an exceptionally large parameter range (pressure, voltage, repetition rate). However, for a lamp this range need not be as large as for a pseudospark switch. The repetition rate should be high, and the lamp does not necessarily need to be able to function at low frequency.

The voltage will also be relatively constant at a few kV, and need not be variable over several orders of magnitude like a high-power switch. This setting of boundary conditions should allow some simplification to the design. For instance, at high repetition rates residual ionisation from the preceding discharge will more than likely be available in the hollow-cathode. This is seen to allow triggering even without an auxiliary glow-discharge. A summary of the currently available technology is shown in figure 3.16. The general trend is clear; the higher the pressure, the lower the maximum operating frequency of the system.

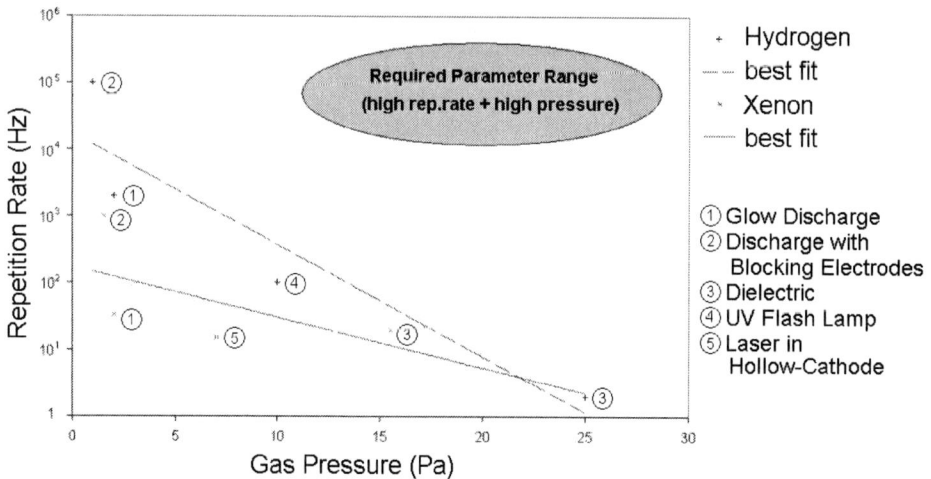

Figure 3.16. Parameter range of currently available methods of triggering a hollow-cathode discharge. The data from hydrogen are taken from literature, whereas those for xenon are the results of measurements made on similar constructions.

Similar set-ups to the ones constructed were used to reach repetition rates of up to 100 kHz, however, not with xenon and not with the energy being dissipated within the electrode system. As the lower line in figure 3.16 shows, much poorer results are obtained with xenon in the same systems. This dissipation of energy leads to increased erosion rates. For the electrical triggers and the flash lamp, used in the same form as for a switch, they suffer from an inherent lifetime problem: The triggers require an insulating surface inside the hollow-cathode to function. In order to increase their life span, this insulator must be moved away from the plasma (if this the cause of the metallisation problem) or replaced by some alternative. This writes off the long-term use of the flash lamp and dielectric triggers. Also, metal surfaces are easier to cool.

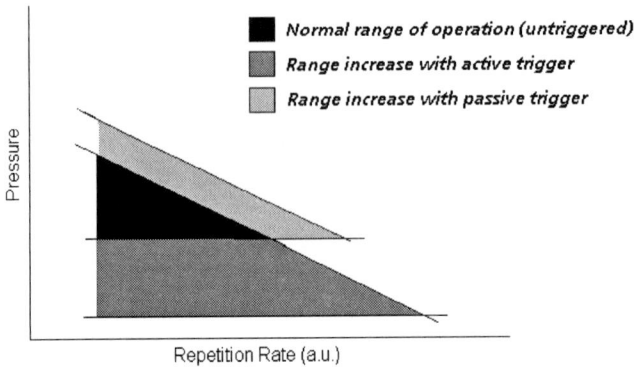

Figure 3.17: Qualitative picture of the available window of operation with the untriggered lamp (black) and with the active trigger (dark grey) and the passive trigger (light grey). Both triggers offer access to higher frequency but in different pressure ranges.

The maximum frequency reached with the untriggered lamp was in the range of 1 kHz. This was achieved by consistently reducing the pressure as the frequency of the lamp was increased to overcome early discharges taking place during the charging phase. A limit is reached when the pressure is so low that the lamp no longer discharges for the first shot. This window of operation is represented by the black triangle in figure 3.17.

Active triggering does improve operation over the non-triggered system at low frequencies, and the possible reduction in pressure increases the EUV signal at the detector because there is less xenon in the optical path to reabsorb the light. Furthermore, at such low pressure (< 0·1 Pa) frequencies of several kHz can be reached, yet there is insufficient xenon to properly form a pinch or radiate EUV yet operation must remain in the lower, dark grey, section of figure 3.17 to prevent extra discharges. In section 2.2.3 the critical particle density was introduced as the condition defining the necessary amount of gas so that the glow-to-arc transition can take place: This can be calculated for our lamp, given a gap separation of 5 mm filled with xenon:

$$n_{cr} = \frac{1}{\sigma_i d} \sqrt{\frac{3m_e}{M_i}} = \frac{1}{10^{-20}.5.10^{-3}} \sqrt{\frac{3.1}{1836.131\cdot 3}} = 0\cdot705.10^{20}\,\mathrm{m}^{-3} \qquad (3.6)$$

where a value for $\sigma_i = 10^{-20}$ has been taken[41] and the mass entered in atomic units. This value of number density corresponds to about 0·3 Pa based on an ideal gas and is in close agreement with the experimentally found pressure at which a pinch is no longer seen using an active trigger. Unfortunately, it is only at these low pressures that the active triggers can function well at high frequencies. For comparison with the switch system, replacing the atomic mass of the xenon (131·3) with that of hydrogen (1) delivers a minimum pressure of 3·4 Pa to form an arc. Figure 3.18 shows the result of plotting critical pressure as a function of the atomic mass. It clearly shows the advantage of using lighter gases if operation at higher pressures is desired.

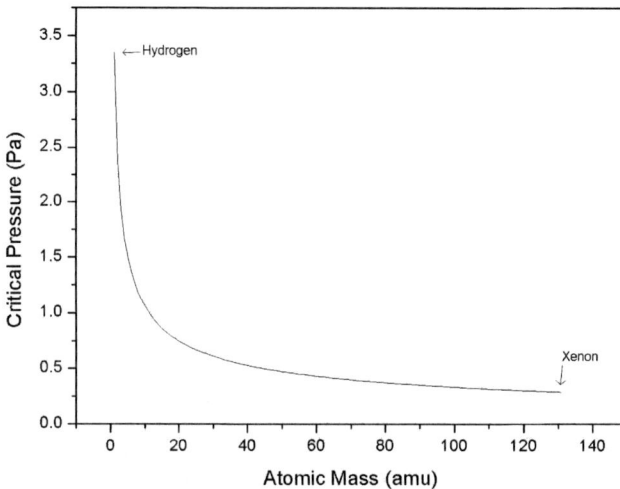

Figure 3.18: **The critical pressure, above which an arc can be created, falls of as the inverse square against the atomic mass of the fill gas in the pseudospark discharge system (calculated here for 3 kV across a 5 mm gap for all masses).**

In the high frequency limit only the passive trigger allows any access to pressures higher than that of the static breakdown and therefore, increased EUV production.

Experimentally, the range of operation (black and light grey in figure 3.17) is smaller than for the active trigger, but there is no necessity for a large pressure or voltage range.

4. The Passive-Triggered Lamp

4.1. Demands on the Trigger for a High-Power Lamp

The triggers tested in same form as for switches were not able to satisfy the demands set out for the lamp in section 1.3, generally because of their inability to simultaneously function at high pressure and high frequency. The EUV output power, initially in the mW range, must be improved by around four or five orders of magnitude and the lifetime from about 10^5 shots up to 10^9. This is obviously not solely the job of the trigger. Other research is going on into lamp design, pinch geometry and radiating media. The trigger should keep the lamp running: stable and contribute considerably to the improvements. It should be able to withstand the power loads for the desired lifetimes even if it is not responsible for these achievements.

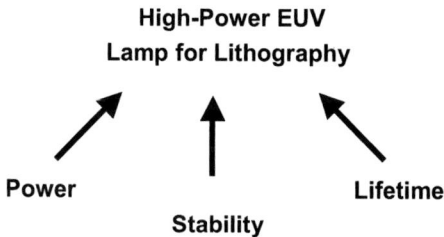

**High-Power EUV
Lamp for Lithography**

Power   **Lifetime**

Stability

Figure 4.1: The key factors which determine the success of a high-powered EUV lamp for lithography are found to be the output power, the stability and the lifetime of the lamp.

Erosion and material build-up on insulators and electrodes and parasitic discharges over the insulator have been identified as the main factors influencing the lifetime of the lamp. Parasitic discharges at higher pressure, frequency limits caused by the use of xenon and efficiency are limiting the power produced by the lamp, as well as the input power itself.

It should be noted that the triggering systems found in literature are for pseudospark switches and not for lamps. In switch systems, the energy stored on the capacitor banks is dissipated in an external load rather than in the electrode system itself, which makes copying the switches directly difficult, due to the much higher thermal load.

4.2. Function of the Trigger

4.2.1. Introduction

The blocking electrode trigger is an auxiliary electrode, often seen in past designs of pseudospark switches.[33] Usually the task of the blocking electrode is the prevention of pre-discharges when the conditions within the system are near to those of self-breakdown, i.e., the limits of the design. When the voltage on the main gap nears the static breakdown potential, the pre-current (charge carriers drifting in the field) is seen on the main gap. This current increases with time as described in the phases in section 2.2 until the lamp fires. Based on this, the intention of the passive trigger is to hold back this pre-current to the main gap. For instance this current can be intercepted to the passive trigger electrode instead of flowing between the cathode and anode of the lamp[38], holding the system in the Townsend phase. Applying a trigger in any of the later phases is unlikely to hold back the discharge, as the electron multiplication and level of ionisation are then too great, so the "point-of-no-return" probably lies at the end of the Townsend phase.

In most switch designs the blocking electrode is found external to the hollow-cathode and the seed electrons are collected via small apertures in the walls of the hollow-cathode.[33] However, it may be introduced directly inside the hollow-cathode, as seen in a few switch designs[30], where a voltage of up to a few hundred volts is applied between the cathode of the switch and the blocking electrode. This type is chosen for the EUV lamp, as it is expected to be more efficient in close proximity to the main gap, and in this case, is the only source of control over the discharge.

The basic principle of the passive system is to hold the device in such a state so that the electrons from the hollow-cathode are extracted directly to the blocking electrode rather than them entering in the main gap. With this in mind, the trigger should be designed as to generate an additional electric field within the hollow-cathode volume (to attract the electrons) without disturbing the charging process of the main gap. Entering a simple lamp geometry into a simulation program, the equipotential lines of the vacuum field can be calculated, the results of which, are shown in figures 4.2 and 4.3 for the cases with and without 300 V applied. The form of the trigger electrode is at this point simply a cylindrical block allowing a large surface area for efficient collection.

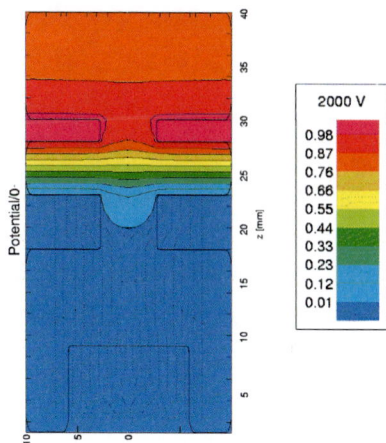

Figure 4.2: Equipotential lines of the vacuum field with the passive trigger inactive.

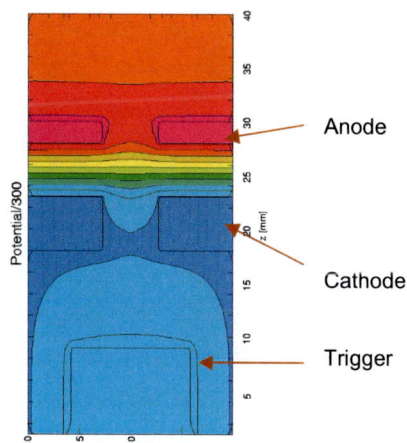

Figure 4.3: Equipotential lines with 300 V present on the blocking electrode

As this output shows, the trigger electrode allows the additional field to completely fill the hollow-cathode, entering slightly into the hole and not affect the field in the main gap. About 95% of the cathode surface is covered by the field from the trigger, which normally contributes electrons. The remainder being the inner surface of the borehole near the gap. So, the function of the trigger will be due to direct charge carrier collection, rather than any other mechanism, such as distortion of the main field preventing the hollow-cathode effect.

In the real system there will of course also be a gas present, which will affect the field over time. Precisely what happens will depend on which gas and pressure is used in the lamp. The pressure which is used, is defined by that required for the lamp to optimally produce EUV. This was initially around 20 Pa (xenon) for the untriggered lamp. At 20 Pa the breakdown voltage for a spark discharge to take place between the trigger and the cathode was found experimentally to be in the range of 2 kV. The trigger is not intended to break down, so this defines the upper voltage limit which can be applied. At voltages a little less than 2 kV a glow discharge forms and if this is also to be avoided, the maximum which can be applied will be in the range of only a few hundred volts as previously found in literature.[38]

The main aim of the passive trigger is in its ability to shift the breakdown voltage (Paschen curve) when the trigger voltage is applied; a higher voltage should be required than available on the main gap for a discharge when the trigger voltage is present.

4.2.2. Triggered Operation

Initially the whole system will be at ground potential and neutral xenon filling the space between the electrodes. Applying now the trigger voltage, without the main voltage present, should show no effect other than a small dark current (experimentally ~ 0·06 mA corresponding to a 33 MΩ resistance between trigger and the rest of the lamp). This allows free electrons and ions to drift and the electrical field to equilibrate, slightly stronger closer to the cathode (due to the lower drift speed of the ions). The cathode-trigger channel doesn't have an electron multiplication allowing any further development.

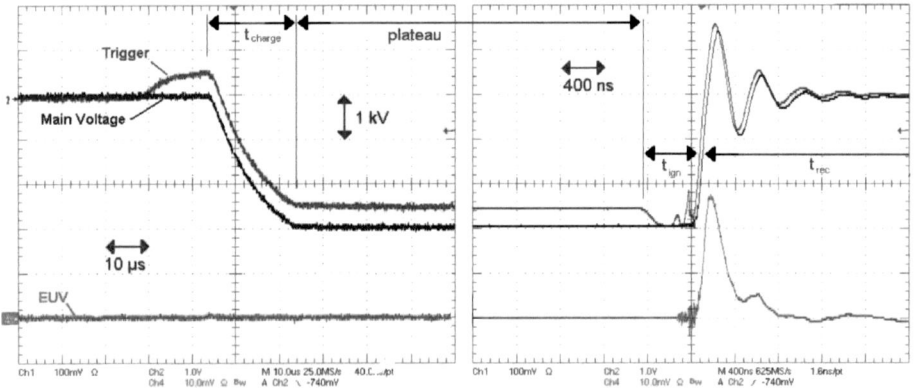

Figure 4.4: **Waveforms of passive triggered operation of the lamp showing the main voltage, the trigger voltage and the EUV signal against time. Note: the time scale is split to show the relevant detail of the two regions.**

We know from earlier that the mean free path of 3 keV electrons is on the order of one or two times the main gap separation and this now sets a boundary condition that these distances should not be available in the hollow-cathode volume. In other words, it should not be too large, in particular in depth, as the trigger is on the back wall.

The usual charging of the main gap can now take place for a time t_{charge}. The cathode is charged negative to a few kV, taking the trigger relative with this. In the main gap a pre-current, made up of electrons and ions will still flow and as before, somewhat more so on the axis due to the holes: Electrons from the gap-side of the hole are still available. What is different is that electrons will not be escaping the cathode volume, so the pre-current will be very much reduced preventing any further development. A Monte Carlo simulation, ran with the addition of a trigger electrode on the rear wall of the hollow-cathode, showed no deviation from the charge densities shown in the pictures in section 2.2 whatsoever (i.e. the results are identical to those of figures 2.5 - 2.7), which leads to the conclusion that the passive trigger has its function after the Townsend phase and that virtual anode may start to enter the hollow-cathode whilst the lamp is held-off by the trigger.

After the required main voltage has been reached, the current from the power supply ceases. During the charging phase, or the hold-off (plateau) phase, some positive space charge could build-up in the cathode chamber due to the increased number of electrons available. If this is the case, it shouldn't be a problem, as it will only increase the current which flows via the trigger and not the level of multiplication.

After a desired time in a plateau phase the trigger voltage is switched off, leaving the entirety of the cathode at equal potential. The time of this removal may be varied and possibly controlled by some external device, e.g. the scanner, when the EUV pulse is required. Obviously, the EUV will not be instantaneously emitted; rather the development of the hollow-cathode discharge will now be resumed as in section 2.2.2. This time delay is referred to as the trigger ignition delay time, t_{ign}.

After the hollow-cathode effect, bottleneck, high current, pinch and emission of radiation have all taken place, as described earlier, the plasma is left to decay. This causes a dead time for the system, referred to as the recovery time, t_{rec}, which must be allowed before the system can be recharged. Recharging immediately would prolong the recovery time by supplying charge carriers to the gap, or if the current the power supply can deliver is large enough, a continuous glow may be produced. The total time needed for one cycle of the lamp can now be defined by:

$$t_{total} = t_{charge} + t_{ign} + t_{rec} \tag{4.1}$$

where any waiting time in a plateau will be included in the charging time. The charging time can be set, the ignition delay time measured and the recovery time determined. The maximum repetition rate possible with the lamp is then simply defined by the inverse:

$$f = \frac{1}{t_{charge} + t_{ign} + t_{rec}} \tag{4.2}$$

There are several parameters which can be used to increase the output power of the lamp. These are: the repetition rate, the efficiency and the input power. As the aim of the project in general is the scaling-up of the lamp to a usable level for production lithography, the reduction of t_{total} is the primary goal as far as triggering is concerned.

4.2.3. Comparison of Possible Electrodes

Four different blocking electrode designs were tested experimentally in the lamp (initially without a window for the light). Each of the four trigger electrodes are shown in figures 4.5 to 4.8. Individual measurements of breakdown voltage and the change in breakdown voltage with respect to the gas pressure (Paschen curves) were recorded and triggered operation of the lamp tested in each case.

In all four cases the ceramic is not in direct sight of the discharge, the electrodes are water-cooled and the main gap is fed by charging the anode positively to 4 kV with the cathode held at ground potential.

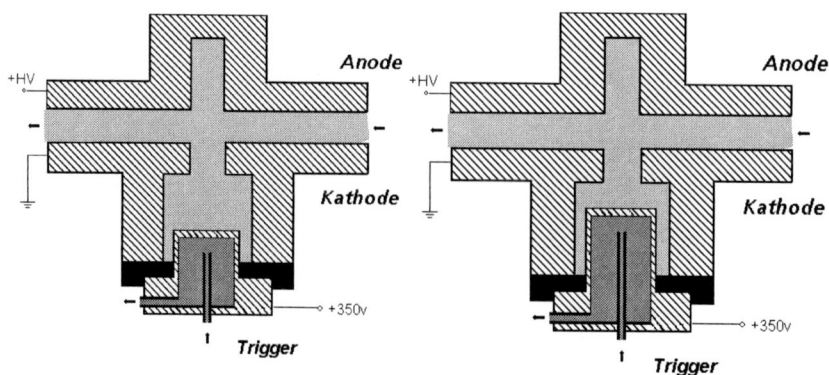

Figure 4.5/4.6: The first trigger electrodes. One short version leaving the geometry of the lamp similar to the untriggered version and a longer one with its surface closer to the discharge region; intended to increase the effect of the trigger.

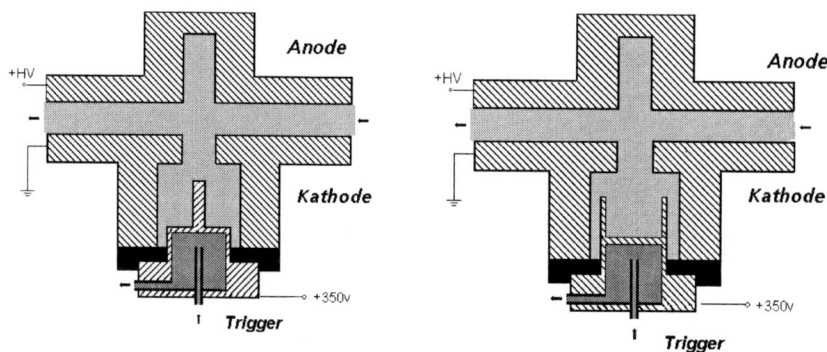

Figure 4.7/4.8: One blocking electrode with a small surface area and one like a cup to allow a much larger surface area for charge carrier collection and in both cases keeping the original hollow-cathode volume.

The first trigger (figure 4.5) had some effect on the discharge but was not able to prevent it taking place and triggered operation was not possible. Only a change between stable and unstable operation was seen as a large jitter in the delay time. The field created by the trigger is too weak close to the hole. The longer the trigger electrode is made, the greater the problem of heat removal becomes. All electrodes were water cooled, yet in the case of stainless steel this is insufficient when only a few mm from the pinch and it is in this range where the highest frequencies were obtained; a large diameter, shallow hollow-cathode of approximately Ø20 x 8 mm. With negative polarity no effect was seen on the breakdown whatsoever up to 500 V. (For the results of larger negative voltages, see active triggering.)

The long version (figure 4.6) was used to inhibit the operation of the lamp. When the trigger voltage (> 250 V) was applied the discharge could be prevented completely. Figure 4.9 shows the voltage profile of the lamp in operation. A control signal from a signal generator starts the main power supply, which charges the capacitor banks to 4 kV (in around 100 µs). While the system is in a plateau at 4 kV the trigger voltage is removed and the system can discharge (after a short delay, $t_{ign} \sim 1$ µs).

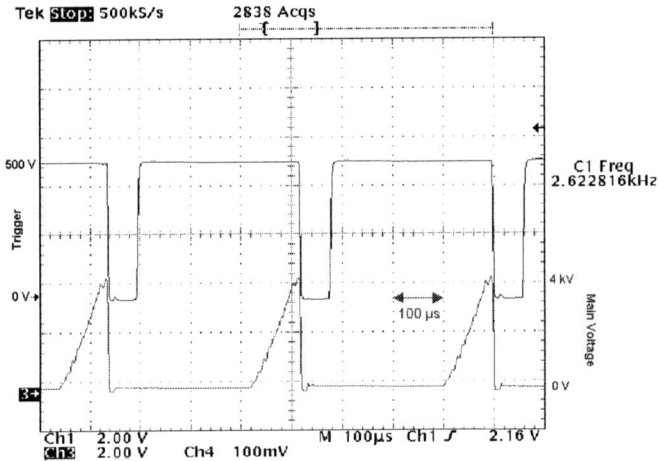

Figure 4.9: **Triggered operation at 2·6 kHz. The anode (lower curve) is charged to +4 kV, which then discharges with the removal of the 500 V trigger (upper curve).**

The trigger is inactive here for only 80 µs, intended to prevent discharges during the next charging phase. A simple experiment, however, shows that this is not optimal. If during triggered operation the main voltage supply is switched off, a bright glow is visible inside the hollow-cathode, so the trigger is hindering the plasma decay process. A reduction in trigger voltage or pressure extinguishes this, yet the more appropriate solution is not to switch the trigger on until the conductivity of the gap has fallen. Figure 4.10 shows how the Paschen curve shifts experimentally as the voltage (constantly applied) on the trigger is increased. By applying a positive potential to the trigger electrode the Paschen curve is clearly shifted to higher voltages. Here again the pressure range for operation is very small and very small changes in pressure correspond to large variation in breakdown potential.

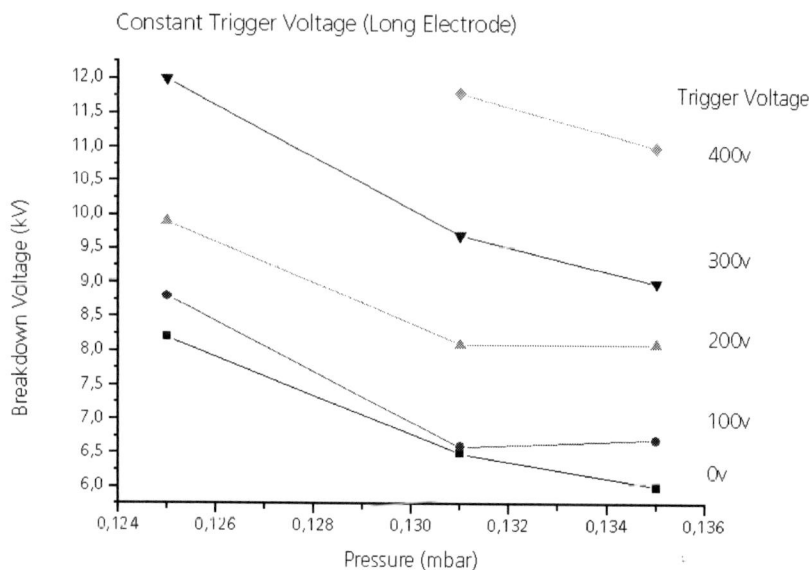

Figure 4.10: **Shift in the Paschen curve for different trigger voltages.**

As expected, a greater voltage is required with the trigger present, due to the electron collection. The frequency limit was around 500 Hz with xenon, with air from the laboratory around 2·6 kHz. Air has a lower average atomic mass explaining the difference in the speed of recovery. In addition, different (much faster) recovery mechanisms exist in oxygen (it being an electronegative gas).

Although the field calculation showed that the effect on the electric field in the main gap would not be so great, the smaller electrode (6 mm diameter) was also used successfully to control the lamp. This electrode had an equally strong effect over the breakdown voltage, its operation very similar to that of the long electrode, showing that such a large area may not be necessary. This electrode did suffer rather badly from lack of cooling.

The cup-shaped electrode should form a stronger field between the trigger and the hollow-cathode, collecting a much larger proportion of the free charge carriers emitted by the cathode surface and experimentally it did show the greatest effect of the four triggers tested for its ability to shift the static breakdown voltage. Its operation was, however, particularly unstable and often discharged early, taking the lamp into a glow-mode. This is most likely linked to its larger capacitance and the speed at which the voltage can be removed.

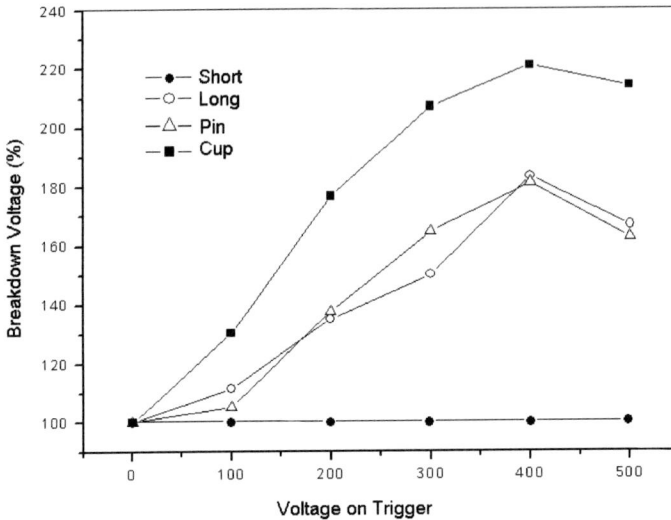

Figure 4.11: **Comparison of the effectiveness of the four trigger electrodes. Normalised in percent relative to the static breakdown voltage with no trigger present. The pressure was kept constant at 2 Pa.**

The three electrodes (the shortest electrode was not suitable for triggering the lamp) all showed a maximum effect on the discharge at around 400 Volts. Over 400 V the charge carriers are being accelerated too much and are thereby causing more ionisation (i.e. it is becoming an active trigger). To compare the effectiveness of the triggers their ability to shift the Paschen curve was measured. Figure 4.11 shows the results of this test normalised to the untriggered breakdown potential of the short electrode. Normalising is necessary as the static breakdown voltages of the untriggered systems are not equal, due to the differing geometries (hollow-cathode volumes), i.e., the change in hollow-cathode volume or also the difference in inductance between the triggers could explain this. For instance, the static breakdown voltage of the smaller trigger is 5 kV whereas the long electrode requires 6 kV.

As the plot shows, the cup electrode has the strongest effect and the short electrode none whatsoever. Between the long and the small electrodes there was very little difference in their effectiveness. The long electrode gave stable operation of the lamp and has fewer problems with cooling compared to the small electrode. The cup electrode showed the steepest curve (optimal for an effective trigger). However, not more than 500 Hz operation could be reached, this was also particularly unstable and pre-ignition was observed often at frequencies as low as 300 Hz, possibly due to plasma inside the cup not being attracted out.

4.2.4. Identification of the Limiting Factor

The reason for not being able to increase the frequency indefinitely comes down, after all hardware has been optimised, to the problem of ionisation left over from the previous shot rather than any other parameter limitation. This problem is particularly important for an active trigger as such a trigger functions by creating ionisation. Below is a screen shot from an oscilloscope of operation just over the frequency limit using the cup-shaped electrode.

Considering basic discharge physics several possibilities should exist for overcoming the problem of residual ionisation and early breakdown of the voltage on the main gap:

Using a mixture of gases containing a small fraction of a lighter element should lead to a reduction in t_{rec}. However, this method is obviously limited to the case when the presence of the other gas starts to hinder the production of EUV. The problem of residual ionisation also still exists for both gases.

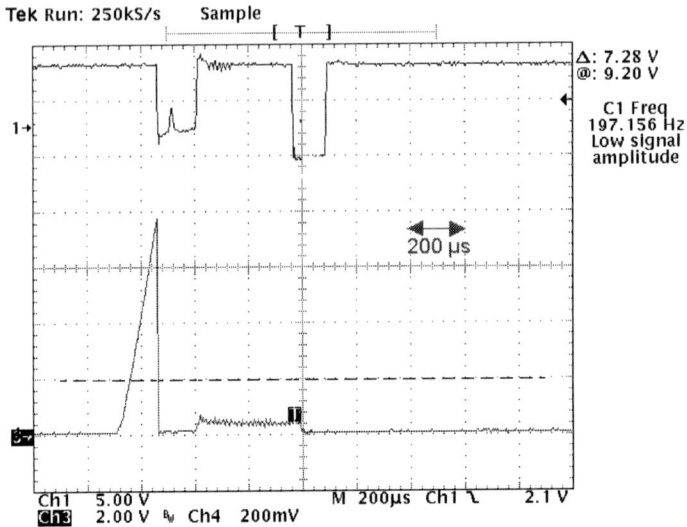

Figure 4.12: Screen shot showing pre-ignition. The positive main voltage (lower curve) discharges before reaching a plateau level. Highly conducting plasma also hold the trigger voltage (upper curve) down for around 150 µs before it returns to its original potential. A glow discharge then burns in the main gap until the supply is removed.

By changing the dimensions of the hollow-cathode the plasma decay may also be enhanced and the probability of pre-discharges reduced. Fundamental theory suggests minimising the volume. Experimentally an optimum was found. However, variation of the hollow-cathode is linked to other parameters, such as the erosion rate of the trigger when brought closer to the pinch, and the speed of the discharge process.

It is already known that by reducing the pressure of the system that the problem will initially be overcome. The voltage required for breakdown will be increased (see also active trigger). This makes early discharge less likely. However, the pressure can only be reduced until there is insufficient gas to ignite the first shot in a burst. This forms a triangular-shaped window of operation for the pressure range, see figure 4.13. This is of course less of a problem in CW operation, where once the lamp has been started; the pressure can slowly be decreased to reach a higher repetition rate. Paschen's law, which is a function of pressure and gap separation, now needs to incorporate another variable; experimentally this would seem to be the repetition rate of the lamp, although more correctly it should be the level of pre-ionisation.

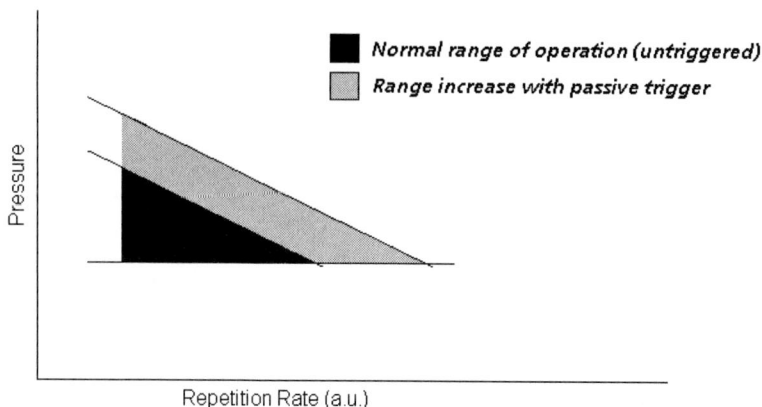

Figure 4.13: The available window of operation with the untriggered lamp (black) and with the passive trigger (grey). As the frequency of operation is increased the window becomes smaller and the lamp very difficult to control. The Paschen curve is a very steep function of pressure, distance *and* level of pre-ionisation.

However, the problem of the much longer delay time incurred by the first pulse in a burst has not been addressed and experimentally it is found that if the pressure is high enough for the first pulse to discharge, then succeeding pulses still have a tendency to breakdown early. In other words, the use of a lower pressure does not solve the problem. Also, it should be noted that the xenon pressure is strongly related to the EUV yield.

The most powerful way of overcoming the residual ionisation problem would be by attempting to charge the lamp and main capacitors on a time scale similar to that of the natural breakdown time of the gap. Current pulse charging technology should allow charging times of the order of a few μs. If this method is used to charge only to the usual voltage, rather than trying to reach higher voltages, no physical process should be able to cause an early discharge. It should then be noted though, that a once a charge is started it cannot be held off indefinitely when run in such a mode.

A popular way of preventing early discharges at higher frequencies (or high pressures) is by actively removing the ionisation in the dead time between the pulses.[33] In pseudospark switches this is often done by the use of blocking electrodes (i.e. by the same method as used here to trigger the lamp). Extending the usage of the passive trigger further, it may also be possible to use the same trigger electrode for blocking the discharge and for clearing the residual ionisation from the system after the main discharge has taken place, i.e. leaving the potential present between the pulses. In a standard switch, decay of the plasma takes place as recombination within the volume and by diffusion to the walls of the system. In a lamp at much higher pressures, methods of actively clearing the ionisation may need to be considered. If the blocking potential is kept on the electrode in the time between the pulses this should extract the charged carriers faster. The size of the current which flows via the blocking electrode, however, should not be too big though, as this could initiate a breakdown in the main gap. So this method is quite limited. Looking at each of the three components, t_{charge}, t_{ign} and t_{rec}, making up the time of the EUV cycle methods of overcoming the limits will be shown for the triggered system.

4.3. Frequency Scaling of the HCT Lamp

4.3.1. Charging Time - t_{charge}

If a voltage is applied to a gap fast enough, a voltage much higher than the static breakdown voltage will be reachable.[55] This is due to the ignition time to breakdown now being long compared to the voltage rise time. Charging in the order of a few μs is necessary. This method has been used in the past in pseudospark switches to overcome residual ionisation problems.[34]

Rapid charging can achieved by using the standard power supply to constantly charge an extra bank of capacitors (continually achieving maximum power from the supply) and then charging the lamp's capacitors when required using an IGBT (Insulated Gate Bipolar

Transistor) switch. The extra capacitors are charged to 4 kV (and the lamp to only 3 kV), which allows very high currents to flow. However only for a short burst until the capacitors are empty. A lot of energy is wasted in the resistance used to limit the current. This would have to be redesigned in any commercial version.

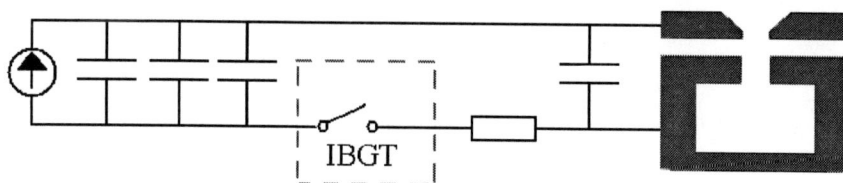

Figure 4.14: Schematic of the "Fast Charging" design. The large capacitor bank charges the lamps capacitors rapidly, allowing the smaller power supply to run continuously at a higher voltage. The trigger is not shown.

Such a system was capable experimentally of charging a lamp fitted with a 1 µF capacitor bank in around 15 µs. The results of applying to the system are displayed in table 1, where as predicted, a gain in frequency is achieved from the reduced charging time, and, a reduction in the calculated recovery time is seen to reduce as a function of the charging time. The ignition delay time does not vary much with the charging time. This was around 10 µs in each case for the lamp tested with the pressure kept constant.

	t_{charge} (µs)		t_{ign} (µs)		t_{rec} (µs)		t_{total} (µs)		f_{max} (Hz)
Low Current	500	+	10	+	740	=	1250	→	800
High Current	250	+	10	+	573	=	833	→	1200
Fast Charging	15	+	10	+	475	=	500	→	2000

Table 4.1: Calculation of the recovery time for three different charging speeds. The maximum frequency was found experimentally in each case.

Attempting to increase the repetition rate beyond the limit leads only to early breakdown of the lamp. The lamp is still being charged slower than the ignition time, but a considerable gain has been made to t_{charge}. Shortening this time more may help reduce the necessary recovery time, but other, cheaper options may be available, such as changing the geometry.

4.3.2. Trigger Delay Time - t_{ign}

How quickly the discharge takes place, t_{ign}, is especially sensitive to the variation in pressure and is the most critical factor concerning the stability of the trigger mechanism. It is also equally a factor of the breakdown voltage, as well as many further, less significant, parameters such as geometry and trigger voltage. The effect is especially sensitive at low pressure (steep Paschen curve). The breakdown delay time, for both passive and active triggers, is defined macroscopically as the time between trigger voltage fall on the electrode and the high-current discharge of the main gap. Microscopically this is the time it takes for the virtual anode to expand and increase its charge density and conductivity of the plasma in the bore hole to a level such that the field strength at the cathode exceeds about 10^{10} V.m^{-1}, enabling cathode spots to form.[36] For useable stable operation t_{ign} is generally much shorter than t_{charge} or t_{rec}, on the order of a few µs or less. Experimentally t_{ign} is also found to fall with the increased repetition rate of the lamp - explained by the excess charge from the previous pulse. The delay was measured for a passive trigger with respect to the gas pressure and is illustrated in figure 4.15. In the case of the active trigger, the curve is

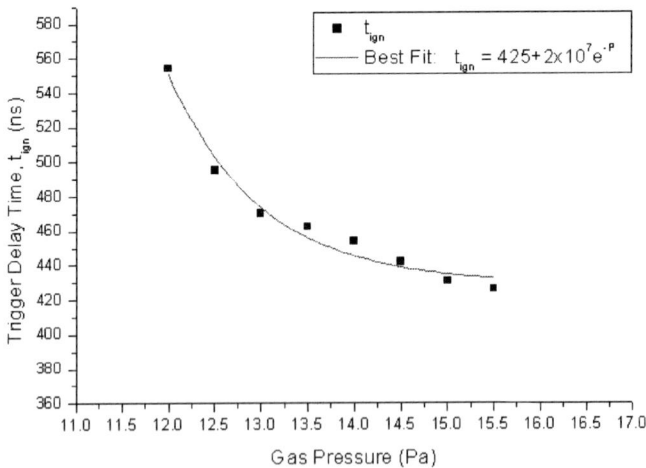

Figure 4.15: Measurement of the delay time between the removal of the trigger voltage and discharge of the main gap for the long electrode at a constant frequency and main voltage in CW operation.

shifted to much lower pressures and then shows also a dependence on the voltage used. Due to the much lower pressure about 8 kV is required on the trigger to operate a lamp with a low enough delay time for scaling up to higher repetition rates. The variation in trigger voltage has only a small effect in comparison to that of the pressure. Higher voltages than this make construction tricky and bring little advantage, as the trigger breaks down before reaching its maximum.

In the range above 15 Pa (this is main voltage dependent) the delay time remains relatively constant, falling only slightly towards higher pressure. At low pressure, a simple exponential decay can be fitted to the data. If the pressure is too low, some pulses have a delay longer then the period of the operation (they are missing, as the trigger switches back on) or no discharge takes place at all and the time tends to infinity. As for the production of EUV, the highest possible pressures prove to be best and this should therefore cause little problem for t_{ign}.

As the main voltage of the lamp remains constant during operation and is defined essentially by the pressure this need not be considered separately. This is of course a different story in the case of the active trigger, as the window of operation for is much larger, albeit towards very low, non EUV-producing pressures, however it was found that by increasing the trigger voltage to 10 kV and above the gas pressure could be reduced to about 0·1 Pa and discharge still initiated. To use this type of operation the dimensions of the lamp would have to be reduced significantly, and the use of a hollow-cathode probably impossible, see section 7.2.2.

The remaining factor which influences the trigger delay time, is not so much the trigger voltage, but rather how fast it is applied or removed. The results of measurements are shown in figure 4.16 where t_{ign} is seen to vary as a function of the trigger fall time, t_f. This is shown for the glow discharge trigger, which is actively switched from ground to 5·5 kV. In the region of interest (i.e. as low as possible) the relationship is relatively linear. Towards very long fall times the operation becomes highly erratic with a high jitter.

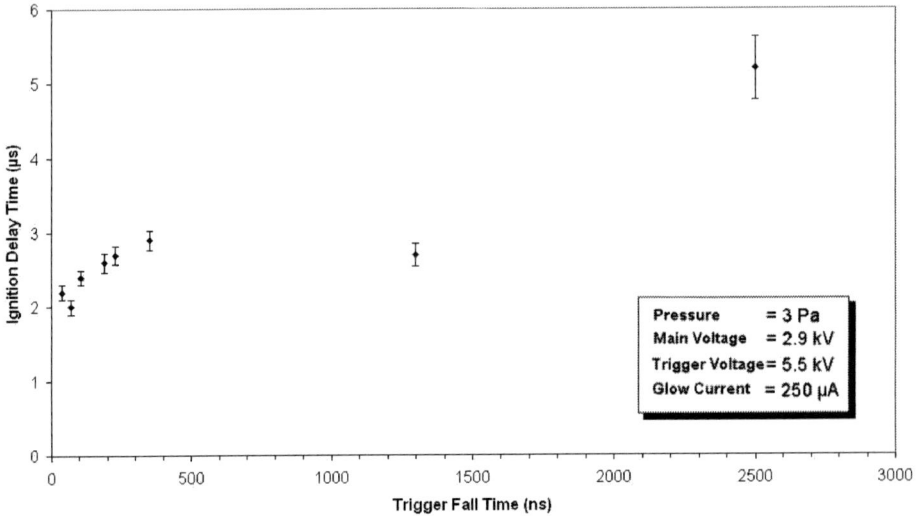

Figure 4.16: Trigger delay time as a function of the fall time. The bars represent the standard deviation of the measurements made (stability).

The result is similar, with much shorter time scales, for the passive trigger, achieved by short-circuiting the electrode to ground. The importance the trigger electrode capacitance and inductance have on the system now becomes evident. The lamp itself has only an inductance of 10 nH or less, so to keep temporal stability and allow access to higher frequencies, the structure of the trigger must be small allowing it to be switched rapidly. Section 6 contains further discussion on stability.

4.3.3. Recovery Time - t_{rec}

Accurate measurements of the lamp data and a comparison of different trigger electrodes and their ability to shift the Paschen curve reveal that by changing the trigger (e.g. 5 mm closer), the Paschen curve (measured with a trigger shorted to the cathode) can be shifted by up to 4 kV. This seems to be related to the hollow-cathode volume rather than the shape of the trigger. Results indicate in general that a larger hollow-cathode requires a lower voltage to achieve breakdown in the main gap.

A larger volume allows more free charge carriers to be available during the Townsend phase, lowering the time required for the field to enter the hollow-cathode. A larger diameter of the hollow-cathode would lead to more interactions from oscillating electrons, speeding up the plasma formation. Also, a longer hollow-cathode in the direction of the pinch would lower the required voltage or time needed for ignition of the main breakdown (Paschen's law).

From this consideration, one could assume that if the hollow-cathode is made very large, the ignition process would be rapid and high frequencies would be achieved. However, t_{ign} is not the limiting process, but rather t_{rec}, the decay of the plasma.[23]

To understand how the geometry of the lamp and its trigger can help increase frequency, plasma decay mechanisms have to be considered. Some details are shown in the appendix, but in general, it can be said the smaller the volume (or closer the ions to the walls), the faster the plasma will decay.

Some results of the calculation in the appendix are shown below in the table for different parts of the lamp. From the table it clear that the plasma in the hollow-cathode is the limiting factor for the decay time.

	Geometry type	Size (mm)	Decay time (μs)
Main Gap (anode - cathode)	Planar	3·0	9
Bore Hole	Cylinder	2·5	11
Hollow-cathode	Cylinder	12·5	270
A new smaller HC	Planar	10	100

Table 4.2: Calculated decay times for different parts of the electrode system.

From this plasma decay consideration, it follows that volume in general must be reduced, especially the hollow-cathode, to achieve higher frequencies. Also the main gap separation, currently 3 mm, could be considered for reduction.

A typical recovery time of 0·5 ms would limit the frequency to 2 kHz, which corresponds well to experimental results. For 10 kHz to be possible, a plasma recovery time, t_{rec}, of less than 100 μs is needed. Fast charging can also allow a higher starting electron density. Using an operating pressure of 20 Pa, the optimum hollow-cathode dimensions were found experimentally to be around 20 mm diameter and 8 mm in depth. This allows the highest operating frequency.

Using the trigger electrode during the recovery time may be a possible method of speeding up the decay process, although the possible voltage, which can be used, is quite limited. Assuming a low enough voltage is used, as not to cause a glow discharge, it should in theory be possible to draw the charged particles to the walls of the system faster than the natural diffusion mechanism. If we assume that a voltage, V, of 100 V is applied to the trigger electrode, then during the recovery time, this would exert a force, $\underline{F}$ on any particle with a charge, q,

$$\underline{F} = \frac{V.q}{d} \qquad (4.3)$$

where d is the width of the hollow-cathode. This corresponds to the acceleration of the charge,

$$\underline{a} = \frac{V.q}{d.m} \qquad (4.4)$$

where m is the mass of the particle. Given that the acceleration is linear, for distances shorter than the mean free path, the time to collect the furthest away electron will then be

$$t = \sqrt{\frac{2d^2 m}{Vq}} = \sqrt{\frac{2.0 \cdot 008^2 . 9 \cdot 11.10^{-31}}{100.1 \cdot 602.10^{-19}}} = 2 \cdot 7\, ns \qquad (4.5)$$

So clearly removing the electrons will not be a problem. For xenon ions the drift velocity must be found, as many collisions will take place in traversing the hollow-cathode. The mean free path of a xenon ions at 15 Pa gas pressure is 0·6 mm, see appendix. The drift velocity, v_d, can be found using equations 4.4 and 4.5 and the acceleration over the mean free path, λ,

$$v_d = \sqrt{\frac{\lambda Vq}{2md}} = \sqrt{\frac{0 \cdot 6.10^{-3}.100.1 \cdot 602.10^{-19}}{2.2 \cdot 18.10^{-25}.0 \cdot 008}} = 1660\, m.s^{-1} \qquad (4.6)$$

which corresponds to a time of 4·8 μs across the hollow-cathode. This is based on singularly charge ions and is significantly less than the current recovery time; so multiply-charged ions will travel faster. In practice, directly after the discharge it is not possible to apply such a voltage, as the conductivity of the residual plasma is so high, such that it effectively short circuits any attempt, cf. figure 4.12, where 350 V could not be applied until after 150 μs, which clearly poses a problem if 10 kHz is to be achieved.

4.3.4. Working Gas

Almost all measurements have been made in 100% xenon, however other gases may be added to the system, either to improve the frequency of the lamp or for purposes of investigation. This often leads to deceptive results - as the operating frequency may be doubled by adding another gas, yet the EUV yield can fall much more. Three measurements were made with: xenon, air and a mixture of the two. The remaining properties of the lamp were unchanged. All three measurements were made at 10 kV breakdown voltage on a passive triggered lamp. The maximum frequencies reached are listed in the following table.

Gas	Max Frequency (kHz)
Air	1·79
Mixture (60:40 Xe/Air)	1·52
Xenon	0·98

A lower pressure is necessary for operation in xenon at the same voltage (its Paschen curve is shifted to the left), explained by xenon's low ionisation potential. The reduction in the particle density may though alone explain the reduction in the maximum possible frequency. The gas flow is lower so less new gas is available. The recovery time for the gas has also to be longer for heavier ions. In the frequency limit the trigger electronics struggle to recover back to the 350 V applied. This will be due to the large amount of plasma remaining in the hollow-cathode shorting out the trigger. As a consequence to these results, an investigation followed into how different positions of the gas inlet affect the function of the lamp.

The trigger mechanism is greatly affected by feeding in small amounts of different gasses through the back of the trigger chamber (active triggered system). A lightweight gas such as helium should allow the ionisation of the plasma to decay very rapidly. Using a very low additional flow of helium (<1 sccm so not to affect the main xenon discharge), this is achieved, yet the lamp triggers only at very high pressures (>25 Pa). The ionisation potential of helium is double that of xenon and therefore less suitable to ignite the glow and form charge carriers for the trigger or hollow-cathode processes. Argon was also tested but the best gas for the active trigger was found to be the xenon with its low ionisation potential and large number of available electrons per atom.

An extra gas inlet near trigger electrode, allows fresh xenon to enter the hollow-cathode region. In previous systems gas could only flow past the hollow-cathode via the main gap. Such a new inlet shortens the time needed for the gas to recover after the discharge and enables a higher operating frequency. However, increasing the flow too far will create a pressure gradient in the borehole region of the system and thus reduce the breakdown voltage, which defines the limit of how much this can help.

Operating with air from the laboratory a frequency of 4 kHz was achieved at 26 Pa with a breakdown voltage of 5·2 kV. This corresponds to double the original frequency possible with the standard gas inlet. If the hollow-cathode is considered as a cylinder of radius and height both 10 mm, then its volume is:

$$V = \pi r_{hc}^2 .h \quad = 3\cdot14 \text{ cm}^3 \qquad (4.7)$$

and then taking the data from the last experiment (4 kHz, 26 Pa) the gas flow, Q_{hc}, in sccm required to entirely refresh the volume between two shots can be found to be:

$$Q_{hc} = \frac{60 P_{lamp}}{P_{atm}} \cdot \frac{V}{t} = 196 \text{ sccm.} \qquad (4.8)$$

Clearly this level of gas flow is not possible without causing a very large pressure gradient at the cathode hole. However, the volume of the borehole though does correspond to the flows used. Taking its diameter of only 6 mm and its depth of 10 mm, this only has a volume of 0.28 cm^3, leading to a required flow, Q_{bh}, of:

$$Q_{bh} = \frac{60 \cdot 29.10^{-2}}{1000} \cdot \frac{0.28}{2.50.10^{-4}} = 17.7 \text{ sccm,} \qquad (4.9)$$

which is in the range used and can therefore explain the improvement found in that the plasma in the bore hole decays faster (see earlier calculation) and this neutral gas is then pushed to the axis of the main discharge where the highest level of ionisation previously was.

4.4. Efficiency Scaling of the HCT Lamp

The conversion efficiency is influenced by many factors: By how much energy remains in the external circuit, resistive losses, current leakage in the plasma outside the radiating volume, out-of-band radiation, ionisation and acceleration of particles. Not all these data are easily accessible by experiment. Impedance matching of the system allows more than 90% of the stored energy to be coupled into the plasma,[16] and although 17% of this does become EUV (9 - 19 nm), a large thermal issue still remains. The following sections will, however, concentrate on those areas which are relevant to the triggered system.

Pressure

As the frequency cannot be increased indefinitely, the energy emitted per shot is the next sensible parameter to be considered. The efficiency to convert electrical energy into EUV radiation is affected by many factors, such as, the geometry of the electrode system, the materials used for construction and the length and shape of the current pulse, but the greatest factor by far is simply the xenon pressure in the lamp. Having found which gas to use and where to allow this to enter the system, the question of which pressure still remains. In early experiments, results hinted that lower pressures allowed more EUV to be collected, and this was the motivation to reach lower pressures through the use of an active trigger. Other than just radiating, xenon also absorbs readily in the EUV region, so with a lower gas pressure the increase in EUV actually comes about due to there being less re-absorption in the xenon atmosphere itself. Also there are less problems of early breakdown allowing access to higher frequencies, and thus more EUV.

So, before a trigger method can be chosen, the conflict must be addressed: xenon is required for creating the pinch and higher pressures lead to higher yield, yet the xenon can reabsorb a lot of the light it produces. Too much gas will cause most of the light to be absorbed and too little will give no pinch in the first place. A maximum must exist where the optimal conditions can be found between the linear increase in EUV production and the exponential decay in the transmission through the xenon, see figure 4.17.

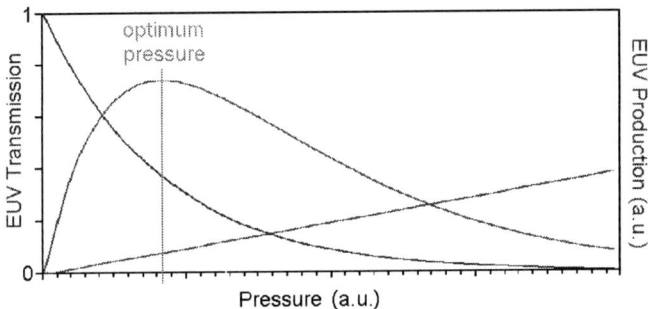

Figure 4.17: Schematic of optimising the conditions for EUV production.

The best pressure for operation of the lamp (pinch position and geometry) was found to be around 20 - 30 Pa, further increase often leads to discharge problems over the insulator. As this pressure lies well to the right of the curves in figure 4.17, a method of overcoming this problem is necessary. The perfect conditions would be the highest possible pressure at the pinch and as little xenon as possible in the optical path to the collector. The addition of a foil-trap, actually intended for the collection of debris, allows a differential pumping stage to be created and this has a lower pressure behind it, and if combined with the introduction of argon, very little absorption will take place in the EUV region. For operation at these higher pressures in the lamp, the passive trigger has to be used.

Voltage

For a fixed size capacitor bank, the use of a higher voltage corresponds to a higher energy. Yet, using a lower main voltage was found to bring a higher efficiency,[46] as this requires a higher pressure and hence more ions available for radiating EUV. Obviously, an increase in the size of the capacitor banks must also take place to keep the input energy constant. The original 20 nF running at around 10 kV stored an energy of up to

$$E = \frac{1}{2}CV^2 = \frac{1}{2}.2.10^{-8}(10^4)^2 = 1 \text{ J} \qquad (4.10)$$

and when replaced with, for example, 1 μF and a reduction of the main voltage is made down to 3 kV, this actually allows an increase of the energy given to the plasma

$$E = \frac{1}{2}.1.10^{-6}(3.10^3)^2 = 4.5 \text{ J}. \qquad (4.11)$$

This reduction in voltage assists the electrical aspects of the lamp, yet the three-fold increase in pressure brings with it a new mechanism; a loss of energy, for which the trigger is found to be responsible. The details of this are discussed in the next section.

4.4.1. Energy Loss

During operation at increased pressures (>15 Pa) and hence lower voltages (<5 kV), problems of energy loss have been found for the passive-triggered system. Often, authors speak of large pre-currents in switch systems; yet never refer to these as being a problem. Here however, the speed of the discharge is vital for EUV production, as the current must be high enough to form a hot pinch and energy lost through any pre-current is undesirable as this is not coupled into the EUV producing pinch.

The loss of energy is seen as a slow start to the discharge, corresponding to a reduction in voltage on the main capacitor banks. This is an undesirable effect when trying to improve efficiency. Lamps and switches, originally designed for operation at higher voltages (~10 kV), must undergo a pressure increase, as known from the Paschen curve, to enable a lower voltage (and keep the delay time to breakdown small). This increase in pressure, however, also decreases the required voltage for glow discharges to burn from the trigger. Two different types of energy loss are identified, corresponding to the cases when the trigger potential is present or after its removal. Both types can cause losses of energy across the main gap (of the stored energy, experimentally this is found to be up to about 25% for former and a couple of percent for the latter). Theory and experiment would both suggest that an ion beam might be responsible for the positive charging of the trigger, and due to the high inductive and resistive connection this charge would not be able to escape fast enough. This is proved not to be the case, but rather two discharge mechanisms, one fed by photons and the other by electron emission, allowing a discharge to burn in the main gap. Physically these are a glow discharge and a spark discharge respectively.

4.4.2. Trigger Glow

If the trigger voltage is too high for the pressure or the repetition rate being driven, then it is seen that a glow discharge starts to burn between the trigger electrode and the cathode. The onset of this glow corresponds to the loss of charge across the main gap. The conditions responsible for this glow are considered further in section 5.3. Experimentally, a current of around 15 mA can be measured flowing to the trigger during this glow phase. This must correspond to a power of 6 W being dissipated inside the hollow-cathode (this was more for larger structured triggers). In the worst case, the entirety of the energy can be considered as being converted to UV radiation, which would release a number of photoelectrons, n_e, from the surface of the system per unit time,

$$n_e = \frac{W_{uv}\gamma\lambda}{hc} = \frac{6 \text{ W}.0 \cdot 1.300.10^{-9} \text{ m}}{6 \cdot 6.10^{-34} \text{ J.s}.3 \cdot 8.10^{8} \text{ m.s}^{-1}} = 10^{18} \text{ s}^{-1} \qquad (4.12)$$

This is the number of electrons leaving a surface with an electron emission coefficient $\gamma = 0 \cdot 1$ and assuming a wavelength $\lambda = 300$ nm. A value of 10^{18} electrons per second corresponds to an initial current, $i_o = 145$ mA, without taking into account any secondary effects or multiplication. With secondary effects, the non-self-sustaining current, i, can be found from

$$i = \frac{i_o e^{\alpha d}}{1 - \gamma \left(e^{\alpha d} - 1\right)} = 172 \text{ mA} \qquad (4.13)$$

where α is the first Townsend coefficient, γ the second and d the gap spacing. Values have been taken $\alpha = 33.3$ m^{-1} for the first coefficient (the inverse of the mean free path for ionisation, see appendix) and $\gamma = 0.018$ for the second (for Xe$^+$ on tungsten electrodes).[47] The narrow gap does not offer much multiplication (simply a factor of 1.2), yet if all the current flows across the main gap, this could theoretically empty the main lamp capacitance of 1 μF in around 17 ms.

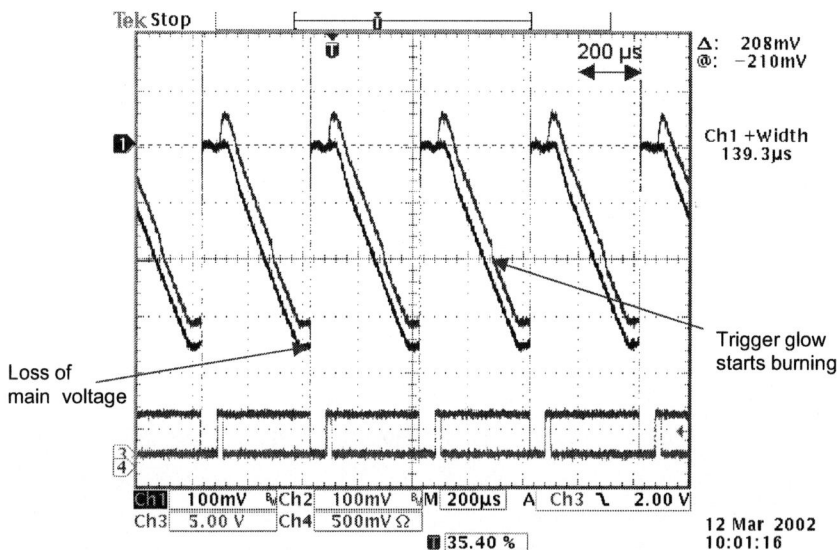

Figure 4.18: Operation showing the trigger-glow and corresponding energy loss in five successive shots. As soon as the power supply stops charging the main voltage starts to fall. The glow discharge here ranges from pulse to pulse between 100 – 400 mA. Channels 3 and 4 are control signals.

If the supply of UV photons is removed, i.e. there is no glow, the current would cease to flow from the capacitor banks,[48] as the mean free path of the electrons is still longer than the gap spacing. This effect of trigger glow does not appear to speed-up or slow-down the main discharge measurably, yet close inspection of the voltage profile of the discharge does show evidence of such a "trigger-glow". Figure 4.18 shows operation with a strong trigger glow and the voltage loss seen on the main gap (a moderate glow would not be not visible on this scale). This operation was made with a high gas flow (60 sccm). After the start of the glow, the trigger voltage is closer to the cathode potential and after the power supply switches off the loss of energy is immediately noticed.

If the pressure were higher still, the trigger would fully discharge and the lamp would then follow suit, as this is then equivalent to the standard trigger method (dropping to cathode potential). If the glow is not present, i.e. at low trigger voltage, the lamp must run close to its self breakdown conditions, or in the case of low pressure, then the ignition process is slow. Both these lead to large jitters, so this glow is actually found to stabilise the lamp's operation if kept small (see also 5.3).

The trigger glow effect has been found experimentally for almost any geometry; for both small electrodes, as equally for large complex-shaped triggers. Experimentally a tendency is found that larger structures initiate a glow much more readily (at lower pressures). MCNP calculations may also verify this effect,[43] yet the results showed large fluctuations and the pressure threshold at which this effect takes place (which is within the lamp's operating range) does not change significantly with the size of the trigger in the simulation.

Generally the trigger glow effect takes place easier with a trigger with a large surface area as the current density need only be very low. Whereas a smaller trigger electrode is found to show a discharge takes place only at very high pressures. As the effectiveness of the trigger is not impeded by the use of smaller electrode, this can successfully replace the large trigger electrodes used previously.

4.4.3. Trigger Discharge

This takes place after the trigger voltage has been removed, i.e. during t_{ign}, however, the ignition process is interrupted. The trigger electrode discharges to ground, reducing the hollow-cathode mechanism and holding the lamp in an enhanced pre-current phase.

Variations made to the passive trigger design show that a discharge resulting in a loss of charge on the main gap can be encountered irrespective of the position of the trigger electrode within the hollow-cathode. A thin ceramic disk, which should be sufficient to block an ion beam, inserted flat in the front face (with highly conductive adhesive for the heat load) does overcome the problem to some extent, but its failure rules out that ion beams could be the cause. The lifetime of the trigger was very short and a conducting layer of metal was deposited on the ceramic after only a few thousand shots.

Whether or not the trigger glow was visible, a positive glow region will be present in front of the trigger electrode surface, created by the electrons leaving the region, ions will have formed (with an greater probability) closer to the electrode having traversed the field from the cathode. After the removal of the trigger voltage the decay of this space charge will be subject to the same processes as the plasma, described in 4.3.3 and the appendix. However, this now coincides with the hollow-cathode effect and the virtual anode entering. If this positive trigger region meets the virtual anode, before the bottleneck phase is reached, then the trigger has a free path to discharge directly to the anode, hindering the further development of the main discharge. This model is shown qualitatively in figure 4.19.

Electrons now trying to traverse the hollow cathode will be "sucked" into the axial low-ohmic path left by the current between the trigger to the main anode, rather than amplifying the plasma density. Experimentally the current continues to ring (in figure 4.20, at about 30 MHz along the axis of the system for around 600 ns) after which the trigger remains at anode potential. The trigger is connected to the cathode, yet due to the high resistance/inductance, it does not recharge on this time scale. How long the system remains in this state varies greatly between lamps, and especially with the pressure, voltage and geometry. It is also combined with very large jitters (tens of μs).

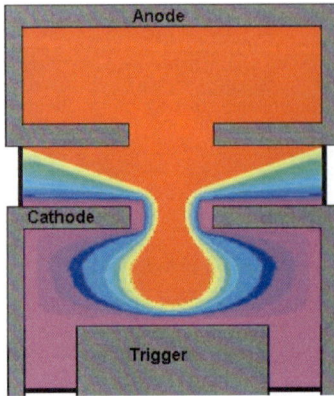

Fig. 4.19a: The start of the hollow cathode effect after the trigger voltage has been removed.

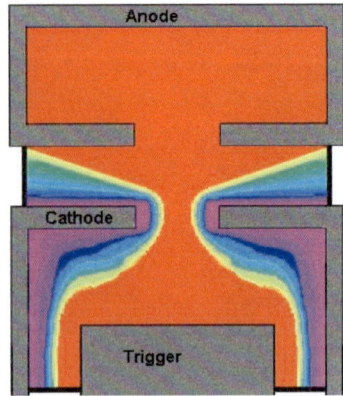

Fig. 4.19b: Suggested form of the field as the trigger breaks down to the virtual anode.

A mechanism such as the photoelectric effect, like for the trigger glow, can now be ruled out for this case. Inspection of the voltage profile reveals that an almost constant current of about 160 A must be flowing across the main gap, in order to discharge the capacitor bank (1 µF) by 30 %. Calculating back using equations 4.12/4.13 from the previous section shows that a UV power of 5·5 kW would be required. Also the question would need to be addressed: Where would this kind of UV power come from? The trigger, this time, is not sitting in a steady-state glow discharge, but instead a rapid (ns) discharge has taken place, which has left behind a mechanism for carrying such a large current, while at the same time hindering the hollow-cathode process.

After the trigger has lost its charge and the ringing has ceased, an amplification of the electron density near the anode will have taken place; as the trigger-anode separation corresponds exactly to one mean free path for ionisation for an electron in xenon at 20 Pa. Given that a current of 160 A is flowing across the gap, with a voltage of 3 kV present, the plasma resistivity in this region can be estimated to be 0·1 Ωm, assuming the current flows through a cylinder 5 mm diameter and 5 mm in length between the two holes. The resistivity of a plasma with an electron density, n_e, can be written as: [58)]

$$\rho = \frac{m_e v_e}{n_e e^2 \lambda} \qquad (4.14)$$

where m_e is the electronic mass, v_e the speed of the electrons, e the electronic charge and λ the mean free path for electrons colliding with the gas. Given a resistivity, ρ, the electron density in this cylinder can be found to be:

$$n_e = \frac{n_{Xe} \sigma}{\rho e^2}\sqrt{2Em_e} = \frac{4 \cdot 8.10^{21}\,\mathrm{m^{-3}} \cdot 3.10^{-20}\,\mathrm{m^2}}{0\cdot 1\,\Omega\mathrm{m}.(1\cdot 6.10^{-19}\,\mathrm{C})^2}\sqrt{2.6\cdot 4.10^{-17}\,\mathrm{J} \cdot 9\cdot 11.10^{-31}\,\mathrm{kg}} = 10^{17}\,\mathrm{m^{-3}} \quad (4.15)$$

where the mean free path is written in terms of the cross-section for ionisation, σ, the neutral particle density, n_{Xe} and the energy of an electron in the 3 kV field, E. This corresponds to an enhanced Townsend pre-current with a glow, but not high enough to form a pinch.[36), 49)] Once initiated the current can sustain itself, as in the earlier stages of the main discharge.

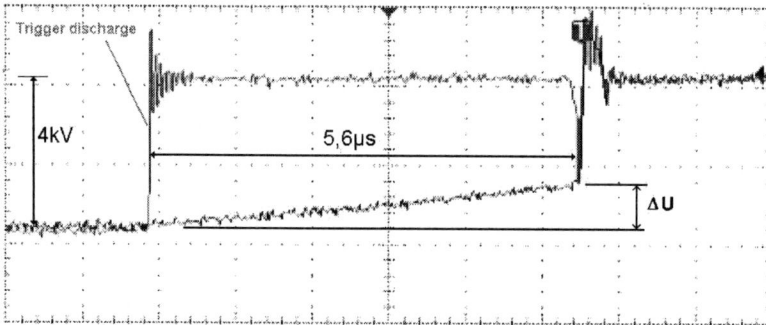

Figure 4.20: The trigger discharge to anode. After the trigger has been removed, energy is being lost from the capacitor banks and is no longer available for EUV production. In this example around 30% is lost (ΔU).

The required extra electron density in the gap region to carry the high current corresponds to an excess charge density of $9.10^{16}\,\mathrm{m^{-3}}$, which in the 5mm x 5mm cylindrical volume equates to about $8\cdot 8.10^9$ electrons. Dividing by the electronic charge ($1\cdot 602.10^{-19}$ C) shows that this corresponds to 1·42 nC; about 2 % of the charge which was available on the trigger electrode, which shows no further mechanism is necessary.

As the plasma near the trigger electrode has no sustaining mechanism this will again be subject to the time scale for decay as in 4.3.3. The trigger breaks back to cathode potential (as the whole hollow-cathode becomes low-Ohmic) and the main discharge process can continue.

4.4.4. Discussion

Given that the sources of the two types of energy loss have been found, it should be possible to design a trigger to minimise the chance of a breakdown to the anode and reduce the intensity of the glow burning in the hollow-cathode. Also, referring back to figure 4.11, the voltage of the passive trigger should be kept low (around 400 V for the dimensions and pressures used here), so that charge carriers are not accelerated, as this is counter effective to the function of the trigger. Glow discharges can still occur if the lamp pressure is too high: a non-triggered system can very easily enter a glow-mode. So the trigger design and operation can never prevent this but should now not to add to such effects. The smaller the surface area of the trigger, the higher the pressure required to ignite a glow. This was shown also by a Mote Carlo simulation.[43] Also experiment agrees, as with the use of a small electrode, like the simulation, the trigger glow is much reduced; it ignites later and the pressure can be increased. Experimentally a loss of about 10 V (10 mJ) is still seen over a plateau of 40 µs. In figure 4.17 this is larger, as this is of the lamp with a much larger surface area trigger. This glow may seem insignificant, but if scaled up to 10 kHz, the trigger glow will be responsible for 100 W of extra heating in the main gap.

By fitting a capacitor (500 nF) between the cathode and the trigger, the trigger discharge problem can be reduced somewhat. The trigger-cathode discharge still takes place, but the ringing is prevented and the waiting time before breaking back is reduced significantly. This is probably explained by the enormous increase in charge density now allowing the trigger to break immediately back to cathode potential. The ringing current is no longer seen. The resistance of the plasma in the gap will be lower still, yet the main discharge takes place much quicker. However, this offers no permanent solution. The electronics take a too large a current and a significant amount of energy is still being wasted across the main gap.

Reducing the size (available surface area) of the trigger lowers the amount of space charge left near the trigger. This reduces the probability of the trigger breaking down to the virtual anode. Also, by substantially reducing the trigger size, the capacitance is small enough so that if a discharge does take place, it will be of an energy low enough not to disturb the hollow-cathode process. Also a change to the dimensions of the system helps greatly. Not having the trigger-gap separation equal precisely to the ionisation mean free path is probably the most significant change to be made. Either raising or lowering the pressure, reduces the problem. A reduction in pressure prevents the discharge, yet leads to a loss in EUV yield.

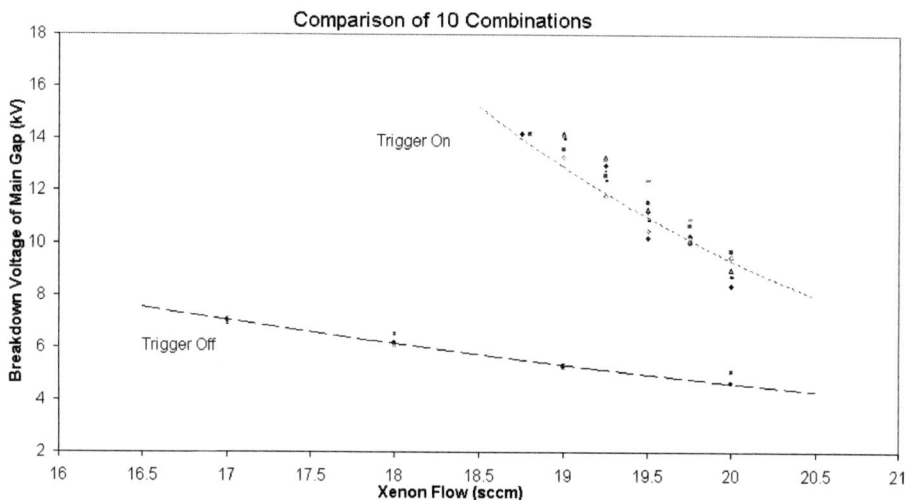

Figure 4.21: Paschen curves for seven different triggers with 500 V applied and three measurements with the trigger inactive. Measurements were made at 100 Hz CW at approx. 10 Pa xenon pressure.

So, the preferred solution is to increase the pressure, so that the electrons from any discharge raises the level of ionisation inside the hollow-cathode, assisting the function of the lamp. Experimentally this is now seen as a small "pre-peak" in the voltage profile and appears not to be a problem to the operation of the lamp.

Without the trigger-glow, the potential of the trigger has a much greater influence on the ignition voltage at any given gas flow (or pressure), whereas the influence of the trigger geometry (the position, size and shape of the trigger in the hollow-cathode) seems to be less important than the other parameters. Figure 4.21 shows the results of several different triggers tested with a constant voltage applied to the trigger and the lamp running in self-breakdown mode. The large separation of the curves (compared with earlier results) shows them very effective at clearing the hollow-cathode of plasma. The close grouping shows the irrelevance of the geometry. This being the case, the trigger shape could be optimised for other criteria, for example its thermal properties, influencing erosion and its lifetime, and more importantly the capacitance (and inductance) of the electrode which has shown to greatly influence the stability.

4.5. Input Power Scaling of the Passive-Triggered System

The input power is clearly related to the output power by the energy stored on the capacitors; in other words, the voltage, the current and the length of the current pulse. However the input power cannot be increased indefinitely, a limit is soon reached and heat removal starts to become a major problem, as does the size and availability of the power supplies.

4.5.1. Cooling

As ever higher powers are pursued, thermal stability of the lamp is of concern, for obvious reasons such as the electrodes melting, but also for the function of a trigger system. It is well-known that systems with hot cathodes have lower breakdown voltages due to the emission of electrons. It is the role of the trigger to help overcome these problems.

A run of 1 hour run made at 3 kW (2·8 J/shot) shows clearly how the original lamp was running at its thermal limit. Four thermal probes were attached to the lamp and temperatures of the capacitors, the air in the lamp cupboard and the front and rear side of the lamp body were monitored. These data are plotted against time in figure 4.22.

Measurements of the cooling water temperatures allow the power distribution in the lamp to be found, see table 3. As can be seen, around 500 W remains unaccounted for; and this is probably lost to the surroundings, as the measurement chamber on the front of the lamps are always found to be about 80 °C. An estimate would suggest this corresponds to around 200 W lost to the air in the laboratory.

	6 min	15 min	24 min	56 min	Average
Anode	800 W	650 W	640 W	780 W	718 W
Cathode	1540 W	1520 W	1500 W	1500 W	1515 W
Trigger	210 W	180 W	150 W	180 W	180 W
Total	2550 W	2350 W	2290 W	2460 W	2413 W

Table 3.   The average heat being cooled away in the three different water channels.

The temperature of the capacitors was a cause for concern. Their temperature (in figure 4.22) shows no sign of ever reaching a maximum. Compressed air used during the measurement as cooling, had a significant effect to the rate of temperature rise, yet offered no permanent solution. A passive capacitor cooling system needs to be used, allowing more air to flow around them over a large area, yet the low-inductive connections must be kept, as this affects the EUV yield. The trigger, however, is much more sensitive to inductance, and therefore cannot be used significantly for heat removal.

In our case, the air in the lamp cupboard needs also be actively cooled by a few degrees and the electrodes designed to increase heat removal. Improved geometry and cooling of the system by using parts from tungsten instead of molybdenum; an increase in the thermal conductivity of about 25 % is achieved (κ_W = 1·74 W.cm^{-1}.K^{-1}, κ_{Mo} = 1·38 W.cm^{-1}.K^{-1}). Together with different capacitors, powers of 4·5 J/shot are now achievable with lifetimes over 10^8 shots. The main conclusion to be drawn is: if any further increase is obtained, from frequency scaling, for which triggering can help, this must be complimented by contributions from the efficiency to produce EUV so that less heat is left in the system.

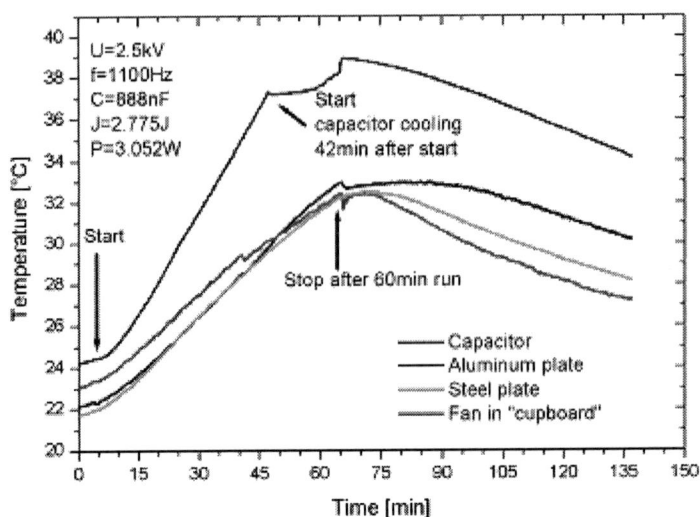

Figure 4.22: Temperature profile against time; heating and cooling. After 42 minutes extra capacitor cooling was employed to avoid their maximum of 40 °C being exceeded. Above this temperature the capacitors would be damaged internally.

5. Lifetime

While HCT lamps are stated to be low debris devices, as the power is scaled-up this does start to play a role in both the lifetime of the electrodes and maybe more importantly to the optics attached to the device. If the lamps are to be used commercially, the electrodes should have a minimum lifetime on the order of 10^9 shots and a frequency of 10 kHz. This corresponds to 27 hours of CW operation.

The relationship between power level and the damage incurred by the electrodes is at first not obvious, but from in-depth investigations, the level of damage appears to be more dependent on the level of energy given per shot rather than the average power level, thought initially to define the thermal limit. Below 3 J/shot there is much less damage to the electrodes compared with those tests over 3 J/shot; whether at high or low frequency. This leads to the need for a higher frequency lamp and a trigger which can survive the harsh conditions.

5.1. Metallisation

The passive trigger, having the form of a small pin, gives very similar Paschen curves to previous ring shaped triggers, confirming that the geometry of the trigger has only a small effect on its effectiveness to collect free charge carriers. Stable operation (without energy losses) is possible as low as 3·5 kV with an upper frequency limit of about 2 kHz. Its small size allows the original hollow-cathode to be as unchanged as possible. At the high input powers now required, deposition of electrode material (mostly eroded from the cathode) is becoming a major problem for such a trigger design. The source of the problem is quickly identified as the high temperature and current. The use of a small surface area electrode allows the thermal load on the trigger to be minimised and also avoids the trigger-glow problem. However, insulators and other surfaces are being covered in molybdenum or tungsten from the cathode.

With the trigger submerged in the wall of the cathode, it is out of reach of the metal vapour. Such a construction shows no signs of trigger glow or discharges and performs better than the previous triggers in both lifetime and frequency.

As the power is increased the lifetime of the trigger becomes ever shorter and eventually a short circuit forms over the surface of the ceramic. Reducing the temperature of the pinch is obviously not an option (see 2.3.4), nor is moving the electrodes much further away from the pinch. Electrodes made from tungsten (or other high melting point material) can only help so much and this material is being deposited on the trigger. The key lies with

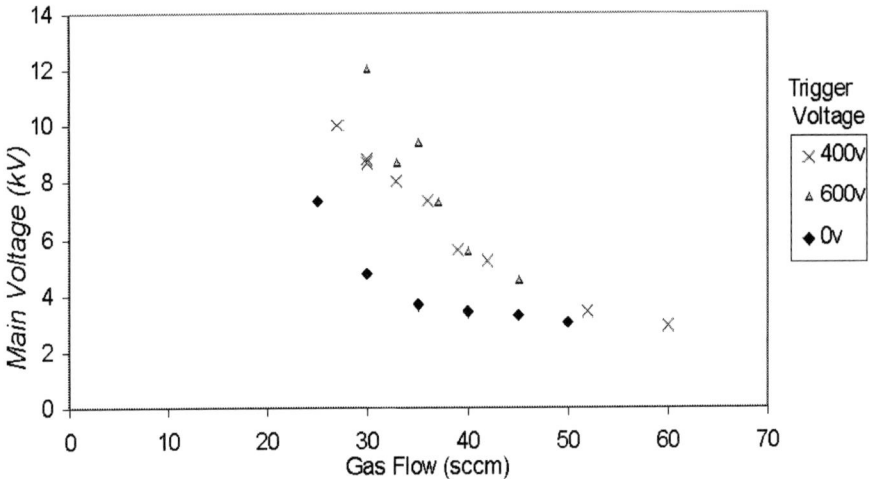

Figure 5.1: **Paschen curves for the submerged trigger. The breakdown voltage is clearly increased by the application of a trigger voltage. The difference between 400 and 600 volts is not as great as for the previous triggers, however at lower pressures the curve is shifted from 5 to 12 kV allowing high stability.**

the efficiency, as the input power cannot continually be pushed up. As far as the trigger is concerned, it has to withstand the power to which it is subjected and designs with shielded electrodes succeed in this.

A second version of the lamp, with a small hollow-cathode, was made, consisting of a small submerged electrode in the rear wall of the hollow-cathode instead of protruding in the centre. This design helps very much towards the metallisation problem on the ceramic. Deposition down a narrow gap to the insulator stops after a few millimetres. The trigger also has no less effect even though it is practically hidden, see Paschen curve in figure 5.1. The lifetime is now limited by the electrode growing in size and short-circuiting the gap.

5.2. Increasing the Lifetime of the Trigger

The lifetime of the trigger unit, whether active or passive, is limited by material debris. This may be material from the trigger itself being re-deposited locally, caused by the intense thermal load or material eroded from elsewhere in the lamp. Tests, carried out using different materials for different parts of the construction and operating the lamp in different ways, show that the source of the metallisation, unsurprisingly, to be the main discharge. Also some contribution from the 8 kV pulses in the case of the of the active trigger was identified. The glow discharge, even running continuously, was found not to contribute measurably to the problem.

In general it is material from the cathode which is found deposited around the lamp. By hiding the ceramic away from the plasma, the lifetime of the active trigger could be increased significantly (the ceramic was never replaced; after many millions of shots). The problems relevant to the passive trigger are discussed in the next section.

5.2.1. On or Off-Axis

The bore hole and rear wall of the hollow-cathode appear to take most of the thermal load and measurements of temperature rises confirm that more power is incident on axis. Around 3 % of the total power emitted by the pinch is radiated into 1 % of the solid angle ($0\cdot13$ sr/4π.sr). The anode suffers also from erosion, but at a much lower rate than the cathode. The material removed is deposited all around the lamp, back on the electrodes themselves, very noticeably on the insulator and on the windows of the vacuum system. Other than the obvious concern for the lifetime of the optical system in a commercial lamp, the electrode, which forms the passive trigger, ceases to function after about 35 million shots if positioned in the centre. A voltage profile of the degraded operation is shown in figure 5.2.

After this time the electrode grows in size due to metallic build-up and consequently short circuits to the cathode. To be successful a trigger must at least match the lifetime of the lamp, i.e. over 10^8 shots and must be able to run at a frequency of several kilohertz with xenon.

The standard electrode used until now has a submerged ceramic insulator to avoid it becoming metallised. Small holes around the base also to allow fresh xenon to be fed into the hollow-cathode reducing the plasma recovery time (see 4.3.4). If the gas is fed in too fast close to the trigger this is found to reduce the voltage required to cause discharges. Hence a large slit with a low flow is preferable.

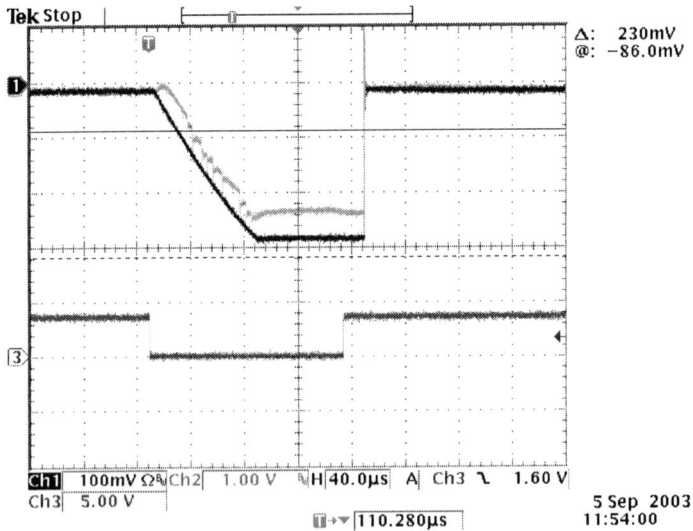

Figure 5.2: Operation with the 1 mm trigger after 33 million shots. Ch 1. - main voltage. Ch 2. - Trigger. Ch 3. - the control signal for the power supply.

The use of a wide hole (several mm) results in limiting the maximum repetition rate to around 600 Hz due to discharges left burning in the hole. The use of a smaller trigger does allow operation at 2 kHz for about 33 million shots, after which the trigger discharges during charging to the cathode. This is similar to the usual lifetime, and is therefore no real benefit.

A trigger with a very sharp point was also tested; its strong field initially allowed 3 kHz operation with no problem, yet after half a million shots stable operation was only found below 1·5 kHz. A small ball of metal had formed on the point of the trigger electrode disrupting its function.

As the build-up of molybdenum (or tungsten) is always concentrated on the axis, where the highest power is incident, shifting the trigger off-axis, combined with the addition of more triggers allows the lifetime to be significantly increased. Having moved away from the problem and reducing the current each individual trigger has to carry, the level of ionisation directly in front of each trigger is reduced.

Experiments on earlier (low power) lamps with off-axis triggers showed no benefit (or hindrance) to the operation. However, at that time the lifetime was not considered critical due to the low power and negligible erosion rates.

A trigger with three electrodes was tested. It is already known that a trigger with a large surface area leads to energy loss from the main gap, so small electrodes identical to the original single one were used. It was thought that one electrode off axis may disturb the pinch or cause uneven erosion of the cathode.

5.2.2. Results

The trigger was put in to operation with a relatively high xenon flow. Inspection of the trigger after several million shots revealed no significant sign of growth or erosion. Usually after this amount of operation the standard single electrode would already show signs of a "mushroom" form. The lamp can successfully run for over 10^7 shots without any intervention and more than 2,5 kHz could be demonstrated at several intervals during the operation for short periods.

After 10^8 shots the trigger electrodes had grown only very slightly and material build-up (mostly molybdenum from the cathode) still took place on the axis which is now further away from the electrodes. Fresh gas was allowed to enter through a central hole in the trigger unit.

The three electrode trigger ran at 1000 Hz for over 10^8 shots giving stable operation with 10,2 W of in-band radiation (in a 2π solid angle at the source). This is a conversion efficiency of 0,27 % from the electrical input power to the capacitors.

6. Stability

The stability of the EUV signal can be considered in three ways: how much is produced; where it is produced and when it is produced. These three, and their influencing factors, will be considered in this section.

A voltage profile of an untriggered lamp is shown in figure 6.1. In operation this forms a saw tooth function. This is due to it being fed by a constant current source. When the voltage is high enough breakdown takes place. The lamp does show stability to both breakdown voltage and the time of breakdown, yet deviations of 10 % in both are not uncommon. The breakdown voltage is dependent on the time of discharge and both are dependent on pressure. The fluctuations will stem from the availability of the breakdown initiating charge carriers. These are from cosmic rays or products of radioactive decay and although their occurrence is common, as is their distribution around the electrode system, their appearance happens at different times and positions. More importantly, is how much voltage is present on the main gap when these initiators occur.

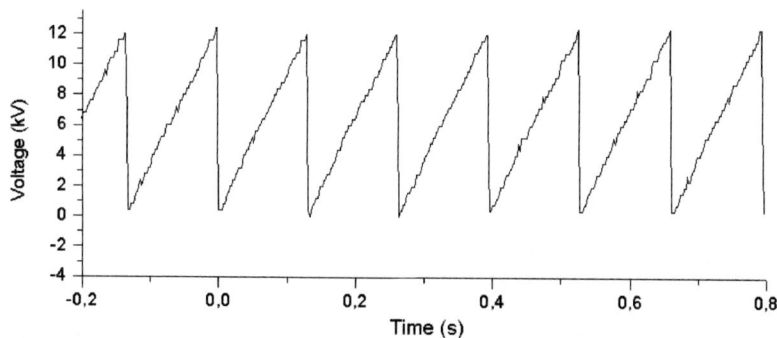

Figure 6.1: Voltage waveform of the untriggered lamp in operation.[50] The lamp charges at a constant current until the breakdown voltage is reached.

The first step on the way to combating this problem is to charge the system to a pre-defined voltage and then switch off the power supply. This allows the fluctuation in the voltage to be eliminated down to the accuracy of the power supply's regulator. To prevent a discharge during the charging phase, the pressure (or voltage) must be reduced slightly, so that operation takes place to the left of the Paschen curve. A pure theorist may now be forgiven for assuming no breakdown will take place. However, the Paschen curve (especially in the case of the pseudospark geometry) is not a thin line. It is in fact a line, with a "grey-area" to its left-hand side, see figure 6.2. In this grey area breakdown will occur. The further to the left one operates, the longer t_{ign}, the delay time to breakdown, will become. Obviously, far enough to the left t_{ign} tends to infinity as does its standard deviation.

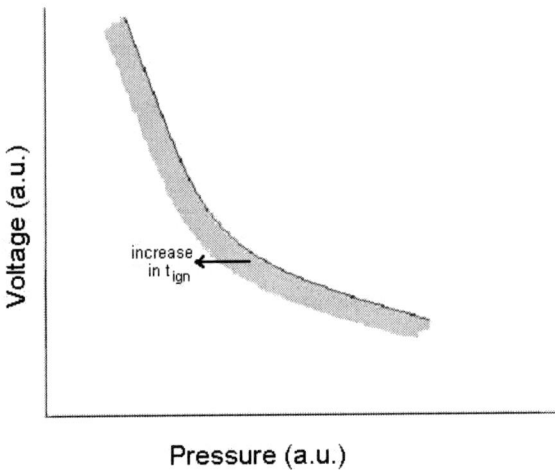

Figure 6.2: Discharge can take place to the left of the Paschen curve, with increased delay time towards lower pressures.

Running the lamp in this plateau mode may eliminate the jitter in amplitude, but it does little for the temporal stability. It is the role of the trigger to initiate the discharge at the desired time and reduce the jitter. The probability of ignition will be described by a normal distribution over time

$$P(t) = \frac{1}{\sqrt{2\pi}} e^{-\frac{1}{2}\left(\frac{t - \overline{t_{ign}}}{\sigma}\right)^2} \qquad (6.1)$$

where σ is the jitter of the delay time, t_{ign}. Reducing the pressure further, until no discharge takes place, allows the use of an active trigger. The length of t_{ign} is then defined by the trigger's ability to inject charge carriers to the system. Of course a jitter does still exist, but this is now smaller. Figure 6.3 shows one of two modes of operation found with the active trigger. This slow trigger mechanism is found during low frequency operation. The trigger injects charge carriers from a glow-discharge and subsequently breaks down directly to the anode (through the cathode hole) simultaneously discharging the lamp.

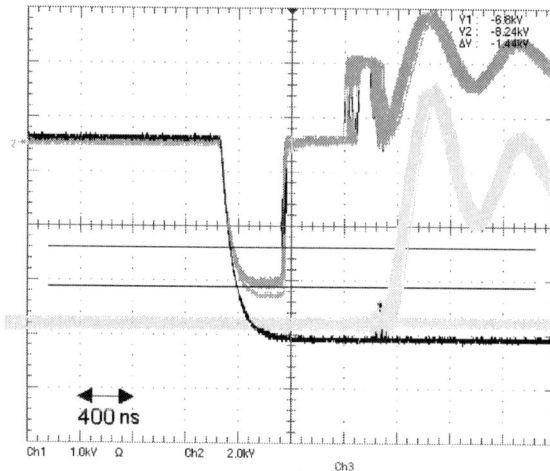

Figure 6.3: Discharge at 400 Hz of the first 43 shots. The first shot discharges late after 18 μs. Ch 1. - main voltage. Ch 2 - active trigger voltage. The darkest line is the trigger voltage of the first pulse.

At high repetition rates (>1100 Hz), the trigger voltage breaks down to the cathode (often before reaching its intended potential, forming a glow). This is explained by the increased level of ionisation allowing a discharge to take place over the short distance in the trigger chamber. A mechanism similar to the slow discharge of the passive trigger then appears to take place. The trigger discharges first to anode potential, however the lamp discharges after such a short time that no energy loss is evident, see the oscilloscope screen shot in figure 6.4. This mechanism is much faster, with very little jitter. As the frequency is increased, the trigger struggles to achieve a potential until eventually the lamp no longer triggers.

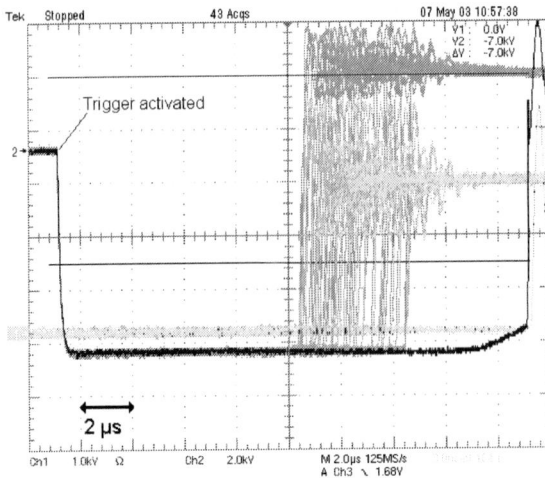

Figure 6.4: The faster, early discharge mode at 1800 Hz. The first shot does not discharge. Ch 1. - main voltage. Ch 2 - active trigger voltage. The dark line is again the trigger voltage of the first pulse.

As the frequency was increased it found, using the glow-discharge trigger, that the glow current was no longer necessary. The lamp would still trigger if the glow was switched off. This is as the residual ionisation of the previous shot has reached a level similar to that created by the trigger. Without the glow however, the first shot does not always ignite, or very late, leading to further instability problems. Just using the glow for the first shot of a burst ought to solve this.

The electrons, created by the glow-discharge, or left from a previous discharge, are injected in the main gap spacing by the active trigger. These electrons will accelerate towards the anode, ionise some of the xenon (if the distance is longer than the mean free path) and a current should flow across the main gap. However, if the electron injection is removed, this [non-self-sustaining] current is expected to cease flowing.[48] This is not the case, showing the difference in properties of the hollow-cathode geometry over plane geometries. Once the charge carriers enter the gap region the hollow-cathode effect commences. Modifications to the electronics showed, that if a negative voltage (~10 kV) is applied to the active trigger, even for a short time (~400 ns) the discharge of the lamp is still initiated. However this does then lead to a longer delay time.

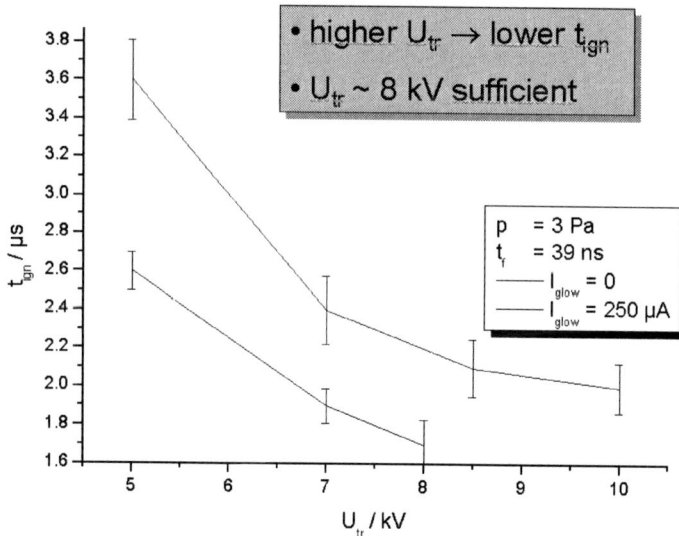

Figure 6.5: Trigger delay time as a function of the active trigger voltage, with (lower curve) and without (upper curve) the glow discharge. The jitter in the delay time is about 350 ns.

Using a passive trigger, the delay times can be made much shorter. This is because operation now takes place to the right of (or above) the Paschen curve. Here however, the delay time is not zero, as the hollow-cathode and bottle neck phases must still take place in a finite time. The removal of the passive trigger can take place very quickly. It must simply be short-circuited to the cathode. Thus, low inductance and capacitance are favourable.

Towards higher frequencies and pressures the passive trigger looses its ability to collect charge carriers. The level of ionisation present in the hollow-cathode will block out the field from the trigger, and increasing the voltage will only lead to (glow) discharges from the trigger, possibly the loss of stored charge and subsequently the lamp discharging early.

Given then that there is plasma present in the hollow-cathode, the question arises: How much plasma can be tolerated so that the lamp will still function correctly. This may be estimated such that; if the Debye length in the plasma generated is less than about one third of the hollow-cathode depth, then the field generated by the trigged will be shielded from the bore-hole region. However, this effect would be self-regulating, because if the field is shielded out, there would be no electrons accelerated towards the trigger to generate further plasma, hence the build-up of the plasma is limited. This leads to a decrease in the trigger's ability to collect electrons and large instabilities should be seen for the trigger delay time with increased pressure and at the frequency limit. The Debye length (see also 2.1) in a plasma is defined by

$$\lambda_D = \frac{1}{e}\sqrt{\frac{\varepsilon_o kT}{2n_e}} \qquad (6.2)$$

So for our hollow-cathode, it would be expected that a Debye length of above 3 mm would be required to allow the field to reach the bore hole. This allows the corresponding maximum allowable charge density to be found:

$$n_e \leq \frac{\varepsilon_o kT}{\lambda_D^2 e^2} = 6.1 \cdot 10^{12} \text{ m}^{-3} \qquad (6.3)$$

Where a plasma temperature of 1 eV has been used. This calculation is valid, however, only for a homogeneous plasma, and this is not the case in the lamp. The mean free path for electrons is greater than the depth of the hollow-cathode, so most will not contribute to the formation of ionisation. Those that do, will form an exponential increase to the charge density. Also, for the region directly before the trigger, this result is clearly too low, as experimentally, a glow is seen. Calculations using a computer simulation based on the Poisson equation, predict that for a glow discharge to be present, the charge density, n_e, must be at a level around 10^{16} m^{-3}. [43), 51)] At this density the Debye length is found to be on the order of 70 µm, which shows that the glow region must be very localised to the trigger electrode, as the lamp is still triggered.

To allow a simple calculation the hollow cathode can be modelled in two regions; one with an electron density of 10^{12} m^{-3}, filling most of the volume, and a second region of 1 mm^3

directly in front of the trigger with a density of 10^{16} m^{-3}. This allows the average value still to be kept.

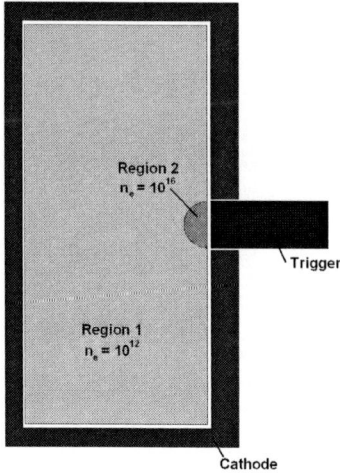

Figure 6.6: Schematic of the two region model for trigger glow in the hollow-cathode.

Under stable operation the glow current to the trigger usually causes a drop in potential from the applied 500 to around 400 V. The remaining 100 V falls across the series resistor (used to protect the electronics). It is therefore assumed that the electrons have a velocity defined by the remaining 400 V electric field. Using again equation 4.15,

$$\rho = \frac{n_{Xe}\sigma}{n_e e^2}\sqrt{2Em_e} \qquad (6.4)$$

the resistivity can be found for the two regions. The majority of the hollow-cathode is found therefore to have a resistivity of

$$\rho_1 = \frac{4 \cdot 8.10^{21}\,\text{m}^{-3} . 3.10^{-20}\,\text{m}^2}{10^{12}\,\text{m}^{-3} . (1 \cdot 6.10^{-19}\,\text{C})^2}\sqrt{2.6 \cdot 4.10^{-17}\,\text{J} \cdot 9 \cdot 11.10^{-31}\,\text{kg}} = 60\,\text{k}\Omega\text{m} \qquad (6.5)$$

and similarly the small region before the trigger is found to have a value of only 6 Ωm, due to the dense plasma.

Assuming now that the glow current flows in series through the two regions predictions can be made for their resistance and hence the maximum allowable glow current. As the resistivity of the glow region is so low in comparison to the rest of the hollow-cathode only region 1 need be considered. This is found to be

$$R_{plasma} = \frac{\rho \cdot l_{hc}}{A_{hc}} = \frac{60k\Omega m \cdot 8mm}{1130mm^2} = 424\,k\Omega \qquad (6.6)$$

where for the length and area, the depth and area of the hollow-cathode have been used respectively. The small area of contact between the two regions is not considered as the cathode surface should be the limiting factor where the resistivity is at its lowest. Simple application of Ohm's law now allows the maximum current to be found under which stable operation should be possible:

$$I = V/R = 400\,V \,/\, 424\,k\Omega \approx 1\,mA \qquad (6.7)$$

Experimentally the lamp can run with higher glow currents of up to a few tens of mA, after which the large temporal instability sets in and the lamp is no longer triggered It can therefore be assumed that the effect of the trigger dies out gradually over this range, starting at a value of around 1mA. To scale-up the lamp further, other methods of clearing charge from the volume between the shots must be employed to keep stability.

The EUV signal with the passive triggered system still has an amplitude stability of ~10 %, even when the voltage is constant, see figure 6.8. The first few pulses of as burst are always smaller, probably due to a heating-up effect (see figure 4.21 and section 3.4.4). The stability of the high voltage is already good with less than 1 % fluctuation. This should be improved further with the addition of the 'Behlke' switch (a MOSFET device designed for pulse to pulse voltage control). Fluctuations of the second half-wave (current reversal) are much greater and correspond to energy left over. This may be dependent on the amplitude of the EUV signal, however, this takes place after the EUV has been emitted and is therefore only relevant for increasing the efficiency if it can be saved or recovered. Figure 6.7 shows the stability of the discharge voltage over a 100 ms burst and the Fourier transform of this signal.

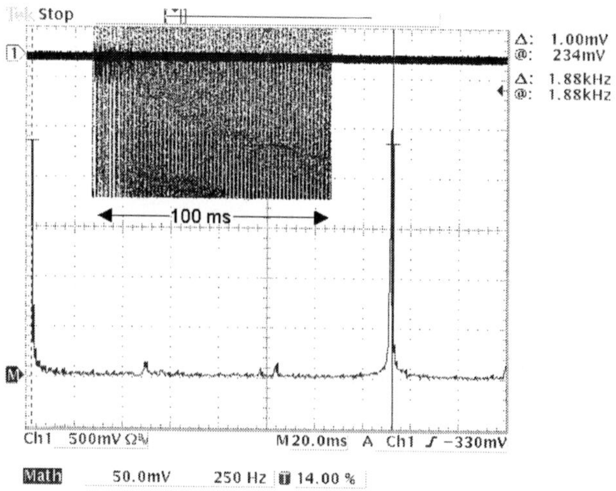

Figure 6.7: A screen shot from the oscilloscope showing a Fourier transform of the discharges over a burst. The upper plot represents the negative high voltage and the lower, the corresponding Fourier transform of the frequencies.

A very clear narrow peak at 1·88 kHz can be seen in the frequency spectrum. This corresponds to the frequency of the operation. Instabilities are very easy to detect using this method. Most often small peaks occur at $^1/_3$ and $^2/_3$ of the working frequency, probably caused by the start-up problems where only every 2nd or 3rd shot discharged correctly. All other noise appears random.

The amplitude is not thought to be influenced greatly by the use of a trigger. This is found to be dependent on the electrode geometry, the main voltage and pressure gradients in the lamp. Likewise, the spatial stability is equally unaffected by the use of the trigger. The position of the pinch is a function of pressure and pressure gradients. Temporal stability, is however, clearly affected by the triggering of the system.

Earlier, the cycle of the lamp was split into three parts: charging, ignition and recovery. The jitter in the charging time is dependent on the external power supply. It is small and is near enough irrelevant when a trigger is used in the system, as the discharge takes place from a plateau. The jitter in the recovery time will be statistical, based on the diffusion of the ions to the walls. The smaller the diffusion time is, the smaller jitter is.

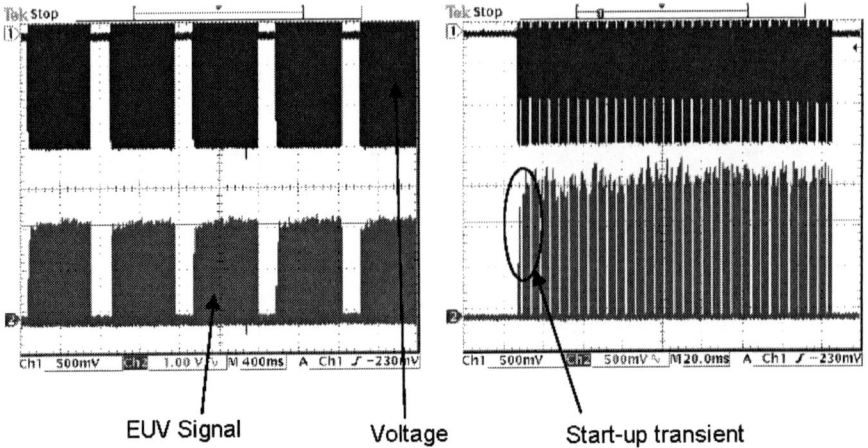

EUV Signal   Voltage   Start-up transient

Figure 6.8: **EUV stability over the burst. An amplitude transient is seen for the first couple of pulses in the burst. For five bursts (left) and magnified (right).**

The greatest parameter affecting the temporal stability of the EUV production is the trigger delay time, t_{ign}. The trigger delay time is defined as the time between applying the trigger pulse and the lamp actually discharging and producing EUV. Desirably this time will be as small as possible, in order to reach higher frequencies. The jitter is defined as the standard deviation of this time. Whilst quite large delay times (several μs) can be tolerated, large jitters cannot. This is strongly related to pressure and the level of residual or pre-ionisation. The question then arises: does the pressure remain really constant? On the macroscopic scale, no variation is seen, but it is known that after the high-current phase, the high-density pinch must rapidly expand, causing a shock wave. If such a wave exists at the time when the next shot is triggered, then there will be a spatial and amplitude dependency on the time of the discharge. Measuring this is difficult, and whether it exists or not, the temporal stability must in any case be maximised in order to improve the lamp's operation. Only then can it be assured that the EUV radiation falls on the correct part of the silicon wafer as it passes through the scanner.

The other start-up problem, when running in a duty cycle, is that the first discharge of the burst has a much longer delay time than the rest in the group. This can be anything from about 50 % longer up to not discharging at all. When the first shot does discharge, it is so close to the second pulse so that the residual-ionisation causes this to pre-discharge. In some instances, the lamp appears to be running at half-frequency, although what actually happens, is that the even numbered pulses all pre-discharge in a glow-mode. This first pulse problem is overcome in continuous operation or very long bursts. Figure 6.9 shows a simplified model, for which the operation is described for a system with a glow-discharge trigger. Real experimental data can be found shown in figure 6.10.

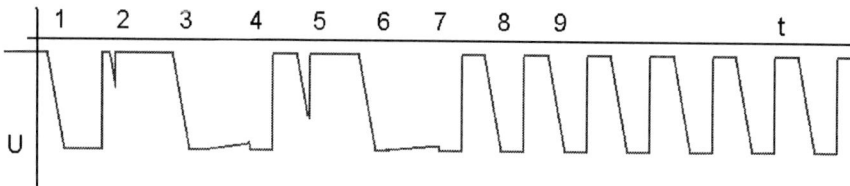

Figure 6.9:   **A simplified model of the first pulse problem showing voltage against time.**

Assuming that the first pulse discharges (if not then there is no operation), this requires a longer delay time than a standard pulse to reach breakdown. This is due to the missing ionisation from the previous pulse (there is no previous pulse or it was too long ago). There is little time available for the plasma to decay before the lamp begins charging the 2nd pulse; this breaks down early. Pulse 3 charges OK as there was ample recovery time in the main gap. In the trigger chamber, however, due its large volume there is still ionisation preventing the same trigger voltage being reached as for the first pulse (the large dark spike in figure 6.10). The lamp fails to discharge now due to the lack of voltage for the trigger discharge and remains at high voltage, discharging as pulse 4 after a delay, similar to pulse 1 but not as long.

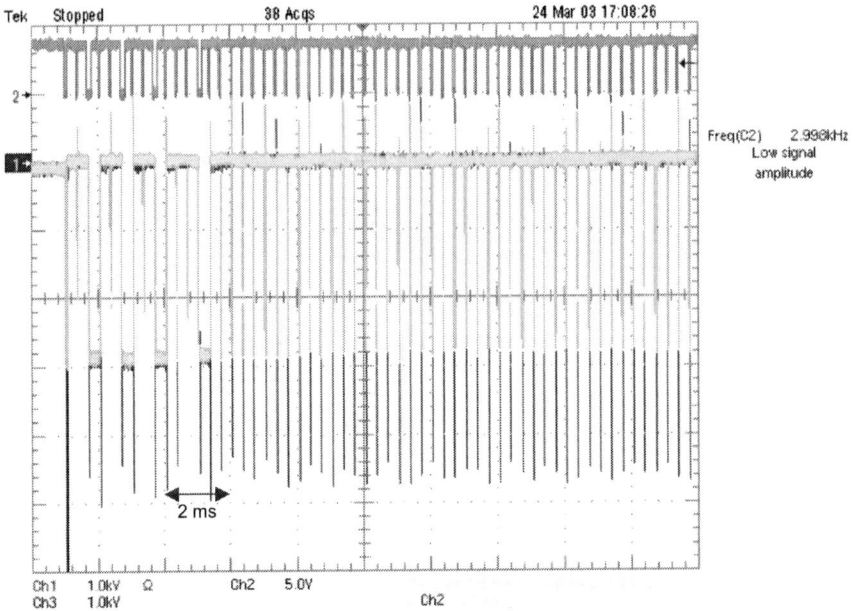

Figure 6.10: Screen shot from the oscilloscope using the active trigger. Only after the 15th pulse does the operation here become stable. Ch 1. - main voltage. Ch 3. - active trigger voltage (mostly covered by Ch 1) and Ch 2. - the control signal.

The ionisation this time is less and pulse 5 reaches a higher voltage than pulse 2 yet still discharges early. The same procedure now takes place, possibly many times, before (as in this example pulse 8) the full voltage can be reached after a full discharge. Consequently (as experiments have shown) a higher frequency can be reached if the lamp is left running continuously. With the passive trigger, only pulses like pulses 1 and 2 are seen (the half or third frequency problem).

One method of overcoming this first pulse problem would be to simulate the residual-ionisation of a previous shot, for instance, with the use of a glow discharge. Another alternative would be the removal of all the residual-ionisation; in the main gap by pre-charging the main capacitors to a few hundred volts. This would clear the gap and allow charging of the second pulse. A greater pressure could then also be achieved. In the case of the active trigger, pre-charging of the trigger for the first pulse also helped reach higher frequencies. Over 4·3 kHz were reached using this method.

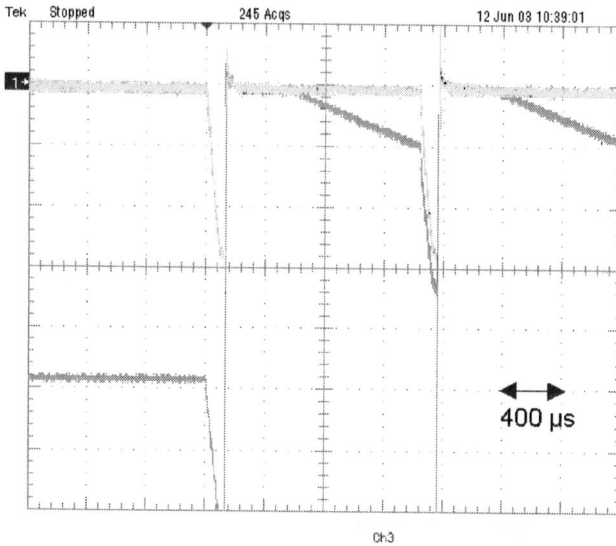

Figure 6.11: 2·5 kHz operation with the active trigger pre-charged to 1000 V. The first pulse is pre-charged much higher to form a glow. This allows the first pulse to discharge correctly. A pressure of 12 Pa at 20 sccm was used.

7. Perspectives

7.1. Active-Passive Trigger

As the frequency is increased, the pressure window for operation becomes ever smaller due to the residual-ionisation of the previous discharge. The passive trigger (or "blocking electrode") collects the residual-ionisation, effectively increasing the required pressure for breakdown, whilst an active trigger injects charge-carriers into the hollow-cathode lowering the required pressure and initiating the discharge. If a system is built, which is less willing to discharge (either lowering the main voltage further or reducing the volume), a combined trigger system could be used. The passive trigger could therefore be used to set the

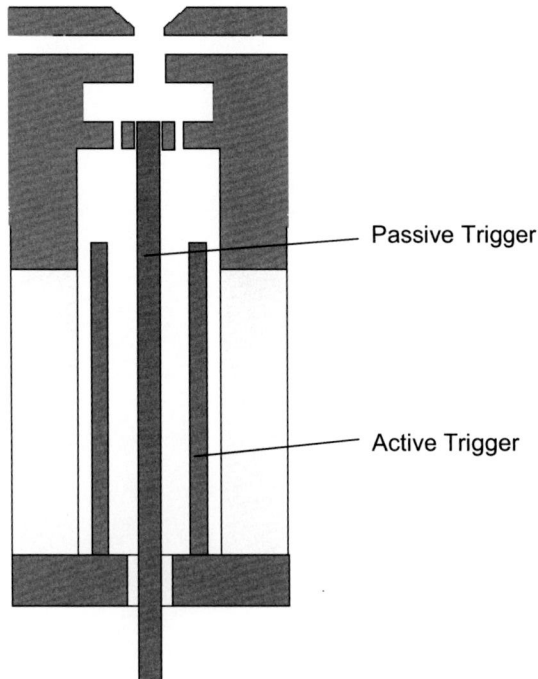

Figure 7.1: Diagram showing the combination of active and passive triggers. The two designs are taken from the individual designs introduced earlier.

breakdown pressure or voltage according to the desired frequency range of operation, while the active trigger should keep the ignition time low. It is expected that the active trigger will be much more powerful than the passive so that the passive may not need to be switched. Below is a possible design for the lamp with a combined active-passive trigger.

This design, combining the best designs (for power scaling) of the active and passive triggers, uses a small trigger (diameter about 3 mm) with a concealed ceramic insulator. This allows operation at frequencies of up to 5 kHz (around 22·5 kW input power). The active part consists again of a long chamber to allow an easy discharge to shorten the delay time to ignition. The overlapping cup-like electrodes form an almost symmetric field for accelerating charge carriers into the hollow-cathode.

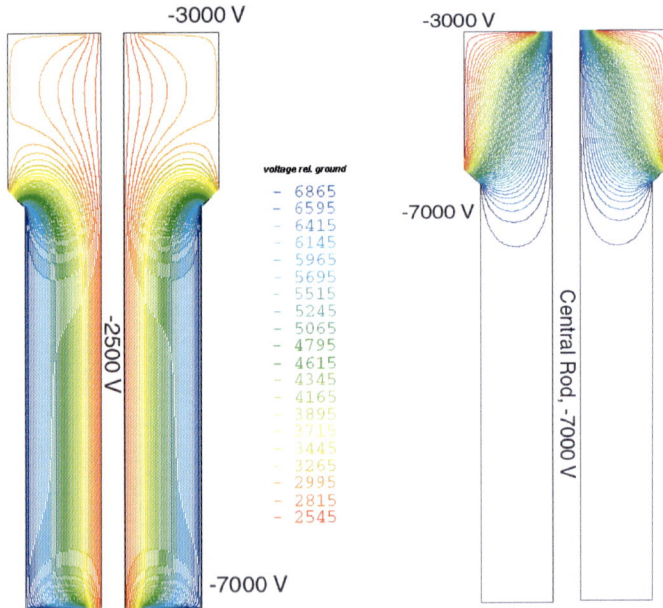

Figure 7.2: a) The trigger active with the electrode left at 500 V.　　**b) The central electrode also switched to -7 kV.**

The passive trigger can be switched from +500 V to match the negative (~7 kV) pulse of the active trigger. Equipotential line plots were made, calculated using software which solves the Poisson equation over a given geometry. These show that if the passive trigger remains at positive potential, the active trigger will only discharge to the passive trigger and probably not assist the hollow-cathode effect. Experimentally the passive trigger does not actually need to be switched, as if it follows capacitively this is also sufficient.

In figure 7.3 the range of operation available to the non-triggered, passive and active triggered lamps is shown. It can therefore be postulated that by combining both a significant gain in frequency could be made. The minimum pressure still remains constant due to the required duty cycle operation (900 ms on, 100 ms off) of the lamp. The first shot is (almost) independent of the frequency, as the previous ionisation has died away.

Results from the active passive combination showed that a much smaller volume is required in order to produce EUV compared to the previous lamps. This is probably due to the higher level of ionisation present, caused by having two triggers.

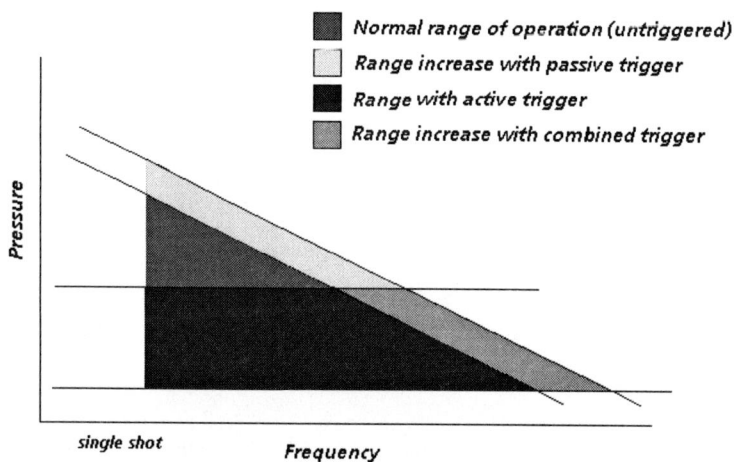

Figure 7.3: Schematic of the available window for different operational modes. It is postulated that by using a combination of active and passive triggering that a much wide pressure window will be available allowing more stable operation due to the lower sensitivity.

Controlling an active-passive triggered lamp also puts much higher demands on the electronics due to the high voltage and necessity of perfect timing of the two trigger systems. This is still a case for further work. Although providing an insulator around the passive electrode, using the experimental set-up of figure 7.1, did improve the field in the rear chamber quite significantly. However the use of such an insulator would reduce the lifetime, see section 5.1, and a different design should be found. This concept in general should function if a lamp can be designed with a reduced volume, sufficient cooling and low capacitive and inductive trigger electrodes.

7.2. The Need for a Hollow Cathode

The main discharge takes place after the hollow-cathode effect has taken place. It takes place towards the end of the bottleneck phase, when conducting plasma fills the cathode volume, much the same as if it were full of metal. So, if the hollow-cathode could be removed at this point, (obviously this is not possible) then the discharge process would continue unaware of its absence. Therefore, if it is somehow possible to simulate the bottleneck phase, a plane cathode, possibly with some borehole or a blind hole, should still function as a lamp: and without a hollow-cathode, much higher repetition rates should be reachable.

If the main voltage is applied to such a plain electrode arrangement, a small pre-current will flow on the axis and the field will notice the hole to some extent, if it is deep enough (Townsend phase). Now, plasma is simply required in the hole and along the axis, to skip the hollow-cathode phase. A concept has been constructed, shown in figure 7.4, which can fulfil this requirement. This design, which is a return to the active trigger, is expected to have a higher lifetime, faster plasma decay and better cooling.

The aim is to simulate the bottleneck effect by producing plasma from the capillary behind the blind hole, generated by the high-voltage pulse in the rear chamber. The lamp in this arrangement functions, and a pinch is visible as usual. An increase in maximum frequency and stability is found over the similar active trigger, yet lower EUV output and performance compared to the passive system, due to its lower operating pressure.

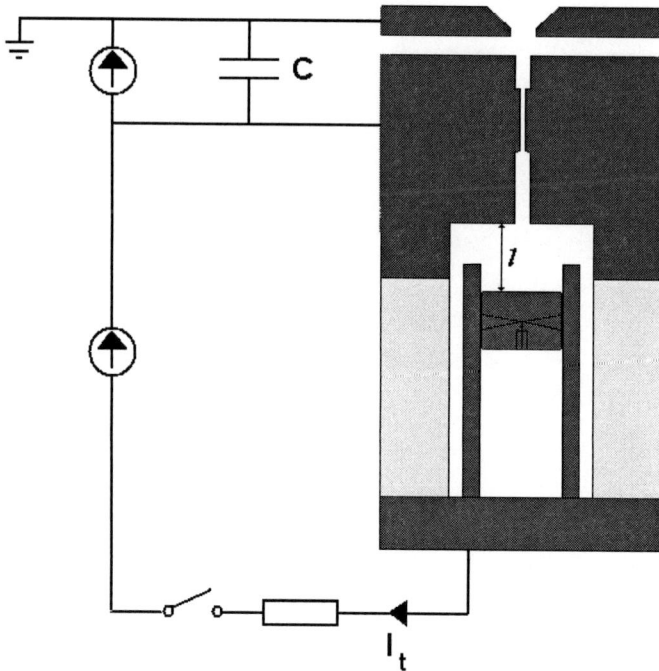

Figure 7.4: The design without a hollow cathode. A piston is used in the trigger chamber to allow measurements at different lengths. The lamp functions, very similar to before with a hollow cathode, and an EUV producing pinch is still visible on the axis.

Measurements were made for different lengths of trigger chamber and the maximum frequency found for each case, these are plotted in figure 7.5. A chamber shorter than 5 mm was not possible, as this led to glow discharges in the system due to the higher pressure required to ignite the trigger. This is of great interest though, as a higher pressure is needed for better EUV production. By reducing the main gap spacing operation in the optimum range (~20 Pa) should be possible.

Using this set-up, two modes of operation were found: the first with the piston close to the cathode, where the trigger discharges simultaneously with the gap and the second with the piston further away from the cathode, where a discharge takes place in the rear chamber.

Operation was not possible at intermediate distances. Here the lamp will not trigger and increasing the pressure brings about only glow discharges in the main gap.

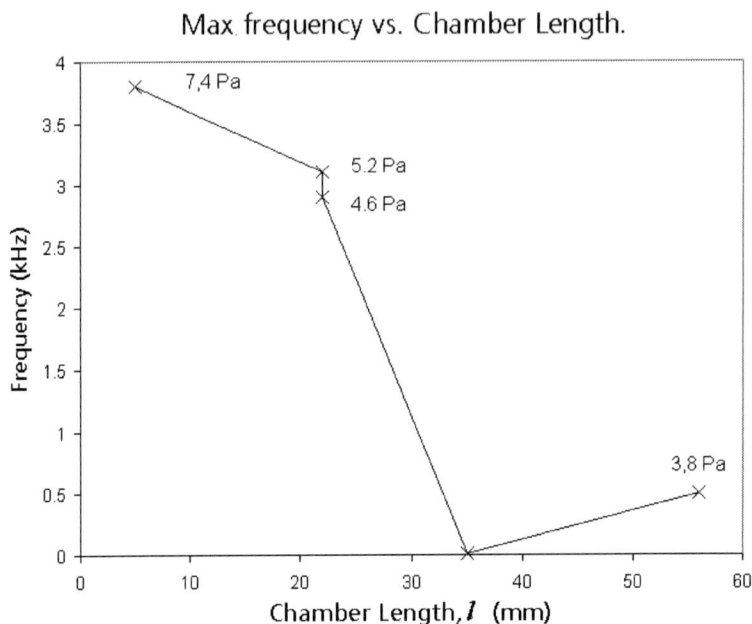

Figure 7.5: The maximum possible frequencies reached with the new design without a hollow cathode. The optimum pressure found for each point is shown. Note: This is the pressure in the main gap, so the trigger chamber may be slightly higher, ~ 0·5 sccm entered via the chamber.

An initial explanation for the two modes is that one mode was where electrons trigger the discharge and xenon plasma the other. Calculations (in the appendix) show that the speeds obtained by the ions lead to a similar time scale to the trigger delay time. Yet the mean free path of the xenon ions shows that it is unlikely for the plasma to traverse the 4 cm capillary to the main gap. Electrons must be responsible for the two modes, yet the question still remains; how? In the close case, those accelerated along the capillary by the high voltage would form ionisation in the cathode hole region (as this happens to correspond to the mean free path for ionisation), however, the discharge is rapid, so further multiplication may be taking place, i.e., the virtual anode enters the capillary. Maybe this still is a hollow-cathode discharge! In the long chamber case, the trigger clearly discharges first to the cathode. Here, secondary electrons from this discharge would be responsible for the same process. The maximum frequency of the lamp appears now only to be limited by the volume of the trigger chamber.

By aiming a laser pulse with an energy of about 0·8 mJ at the bottom of the blind hole (with or without the capillary present), a plasma capable of simulating the bottleneck phase is produced. Again the lamp discharges producing the usual pinch and EUV. Even without the blind hole, the lamp still discharges but this is less stable and a cloudy EUV image is seen. More laser energy is required than with the hollow-cathode present (cf. 3.2.3).

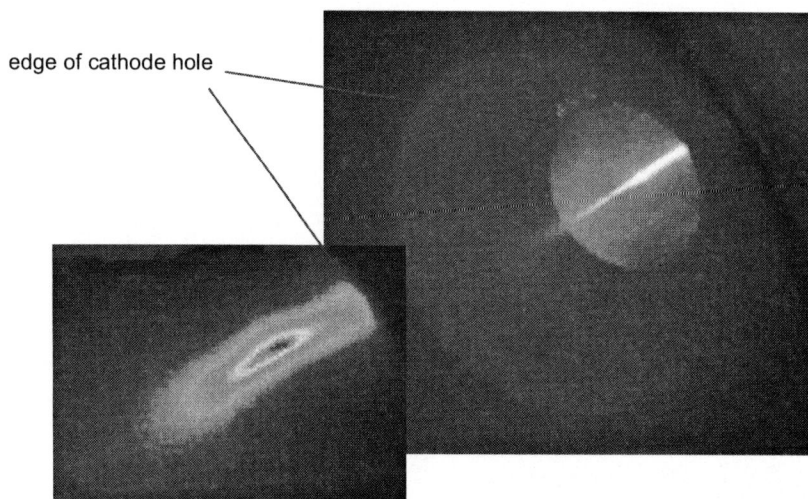

edge of cathode hole

Figure 7.6: **Images of the pinch produced with the laser triggered lamp. In the EUV region with a CCD camera (left) and a similar photograph in the visible region (right).**

The plasma filament is clearly visible with the EUV emission peak localised around the cathode face level. The position of the pinch, with a diameter under 0·5 mm, is almost independent of pressure gradients. Using the capillary as an extra gas inlet up to 7 kHz can be demonstrated producing a pinch smaller than 2 mm in length. At frequencies over 4 kHz the efficiency is found to sink.

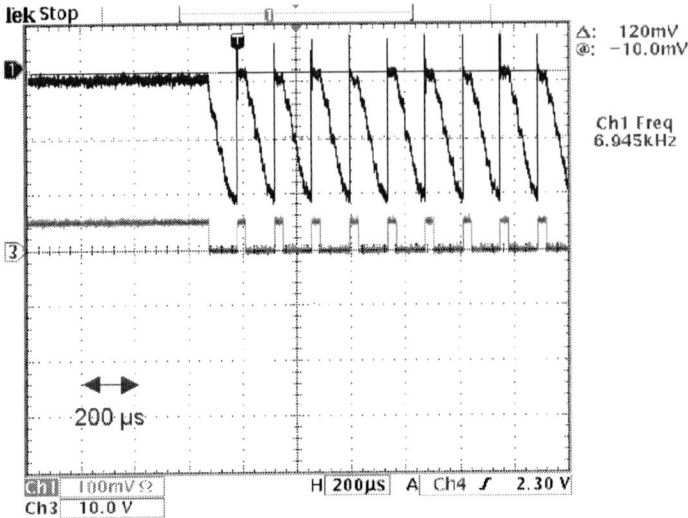

Figure 7.7: Voltage profile of the 7 kHz operation. Ch.1 Main lamp voltage. Ch 3. The signal controlling the laser pulses. The burst is good from the first pulse but only very short due to the enormous thermal load.

Running with the laser has proven that the EUV lamp can produce over 10 W in 2π solid angle (input ~ 17 kW), and that the bottleneck phase can be simulated. However, using the laser, other than for proof of this principle, brings back the problems of difficult construction and high cost. Its lifetime is also very limited and material in the bore hole of the cathode melts, blocking the gas inlet after only 10^6 shots.

Based on these results, and on previous work, such as the three-electrode passive trigger, the knowledge can be combined to build an active-triggered system without a hollow-cathode, intended also to reach the high frequencies, which have been shown possible. Using, for instance, three off-axis capillaries and an arrangement of floating plates in the rear chamber, allowing smaller gap separations, high frequency should be achieved, if the energy of the trigger discharge is high enough (i.e. similar to that of the laser). A sketch of this suggestion is shown in figure 7.8. Having smaller capillaries between the main gap and the trigger chamber allows also different pressures in the two parts of the lamp and near complete independence as far as the plasma decay is concerned.

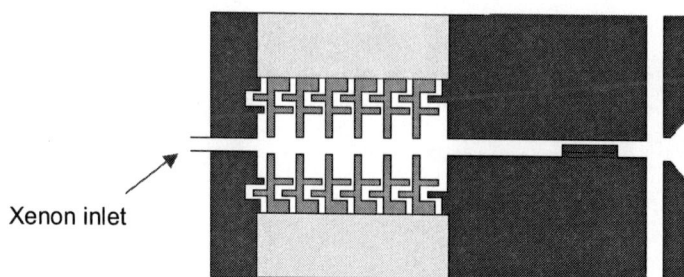

Xenon inlet

Figure 7.8: By filling the volume with disks at floating potentials, electrons can be accelerated in a planar field over a long distance along the axis without the need for one large volume. The design of the trigger now somewhat resembles a particle accelerator.

8. Summary and Conclusions

Without triggering, unstable operation was always seen above 1000 Hz when using xenon in the HCT lamp. Also the breakdown voltage (energy) showed a strong dependence on the gas pressure, gas flow, lamp temperature, and level of ionisation. The aim of this work was to break these dependencies and take control over the operation of the lamp allowing higher power, lifetime and stability.

With passive triggering stable operation above 4 kHz has been demonstrated with only low sensitivity to pressure variations and a pulse timing is possible with a jitter well below 1 µs. This has contributed greatly to the scaling-up of the HCT lamp and has not caused any known reduction to the lifetime of the system. Figure 8.1 shows again the parameter field for the various kinds of trigger mechanism which can be used in an HCT lamp; this time with the addition of the passive trigger, the concept of a lamp without a hollow-cathode and the experimental evidence that this should be possible performed with a laser.

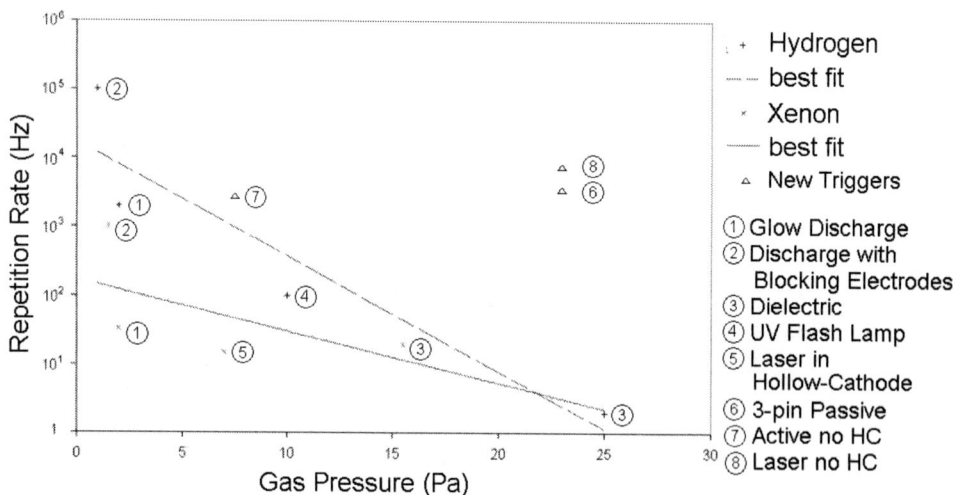

Figure 8.1: Parameter field for different methods of triggering the HCT lamp, now with the addition of the new triggering methods. Passive triggering has allowed the departure from the original trend and operates at high frequency and pressure.

The original untriggered low-power laboratory lamp design could be improved and upgraded to a high-power commercial product as a result of introducing the passive trigger. The current lamp system is controlled via a PLC, has active cooling systems and automatic statistics monitoring and analysis and can handle input powers in excess of 20 kW for over 10^8 shots, producing 66 W of extractable radiation within 2 % of 13·5 nm at the source.

Through the use of an active trigger, evidence was found showing that high frequencies (and powers) are possible with a much larger (yet lower) pressure range, compared to that of the untriggered lamp. These were the conditions believed to be optimal at the time of commencing the investigation, but it is now known that a higher pressure, and therefore a higher xenon particle density, is the necessary condition to gain a higher EUV yield, which together with the use of a non-EUV-absorbing gas in the optical path, delivers the highest EUV power.

The three most significant results of this work are: the finding that the shape of the trigger electrode has very little influence over the change in static breakdown voltage of the lamp; proving that the trigger should be one or more small low inductive points off-axis in the hollow cathode, to achieve a long lifetime whilst avoiding loss of energy and stability; and the conclusion that a limit is reached using the passive trigger in the HCT lamp with xenon.

The detail of the geometry of the trigger electrode has very little effect on the passive trigger's ability to affect the static breakdown voltage of the main gap. Measurements of Paschen curves with the hollow-cathode outer dimensions kept constant, yet with various lengths of trigger showed that its position inside the hollow-cathode is far less important than the size of the volume itself. Taking only perfect measurements without the trigger glow present, i.e., a slightly lower pressure than optimum for EUV production, the breakdown voltage is much more shifted by a 500 V trigger. This shows that in earlier experiments the effect of the trigger must have been shielded by the trigger glow. Testing different triggers (section 4.4.4) showed that only the voltage, or more accurately, the field strength, near the cathode bore hole is the relevant factor.

In order to reach high powers, large robust triggers were assumed to be suitable for the scaling-up of the lamp. It was found that small highly conducting electrodes can withstand far harsher conditions. Moreover, these are much quicker to respond to the electronics due to their low capacitance and inductance and their very small surface area incident to the plasma minimises the cooling requirement and although the erosion and debris show little sign of being beaten by the trigger, the problem can be reduced by the modified geometry and by choice of materials. Parasitic discharges were the limiting factor for the lifetime in the first generation lamps. This problem was identified as surface discharges over the ceramic insulator and was eliminated. The lifetime of both the trigger and the lamp are well on their way to reaching the milestone of 10^9 shots. This has been achieved by moving the trigger off-axis (now with three-electrodes), protecting it from the worst erosion and debris, which helps keep long-term stability. The erosion of the cathode has not prevented but has made less significant to the operation of the trigger and the production of EUV. The debris is still, however, a problem for the optics, which a foil-trap or further modifications have to prevent.

The efficiency to convert electrical energy into EUV radiation increases with pressure, and the loss of energy, which resulted from the increased pressure, has been investigated and the repetition frequency and possible output powers have been improved significantly as a result. However, the conversion efficiency cannot be continually increased by increasing the pressure, as trigger glow problem arises reducing the effect of the trigger. The same applies to scaling-up with an increase in frequency; here the residual ionisation causes the trigger glow. Similarly, an increase in main voltage leads to pre-discharging, and an increase in trigger voltage brings again the trigger glow. The desire to have a higher pressure and higher frequency though, does not mean an end to the lamp. It has been found though, that improvements can be made by varying gas-flows in the lamp and this can allow for different conditions in different parts of the lamp, e.g., a lower pressure near the trigger electrode than in the main gap may produce more EUV without loss of stability. Also, more, and even smaller triggers may hold the key to further scaling. An active triggered system and changes to the geometry could be made to increase the static breakdown voltage, e.g. further reduction of the hollow-cathode or simply bringing the main gap closer together. Efficient high frequency operation is possible at high pressure using the laser trigger and seven kilohertz were demonstrated producing 10 W of in-band EUV, proving that the active trigger should be able to operate a lamp without the hollow-cathode. Combined with blocking electrodes, as in the switch designs, this method could be favourable and new suggestions for an active trigger based on this result have succeeded in preliminary tests.

The maximum repetition rate was found to be determined by the charging time, ignition delay time and recovery time of the plasma, which in turn has been shown to depend on the diffusion rate of the xenon ions.[42] Faster charging and an optimised hollow-cathode have shown that it should be possible to reach repetition rates in excess of 10 kHz if the solutions to the first pulse problem are implemented (section 6).

By charging the lamp on a time scale of a few μs, ten kilohertz can be demonstrated for short bursts with the active trigger. The EUV yield was low and only the first pulse of the burst was actually triggered. The main problem lies in that the conditions required for the first shot of operation requires different conditions to the rest of the burst, which is one of the major influences identified as affecting the stability of operation. Whether active or passively triggered, or even by other methods, if the first pulse does not discharge similar to one in the middle of the operation, start-up transients over many tens of shots can occur. Methods of overcoming this, such as pre-charging or over charging the first pulse are subject to further research.

A combination of passive and active triggering has been introduced, yet not fully tested due to its intricacy. This method may hold the key to reaching stability at even higher frequencies and also more freedom over the operating range. The ignition delay time was found to be a function of trigger fall time and therefore a "quasi" active-passive combination (in which the passive 500 V is switched through to - 500 V) delivers better results, as the fall time of the trigger is being actively shortened. Such a low negative voltage alone is not sufficient to trigger the lamp and if removed again immediately after discharge, no physical reason speaks against this method.

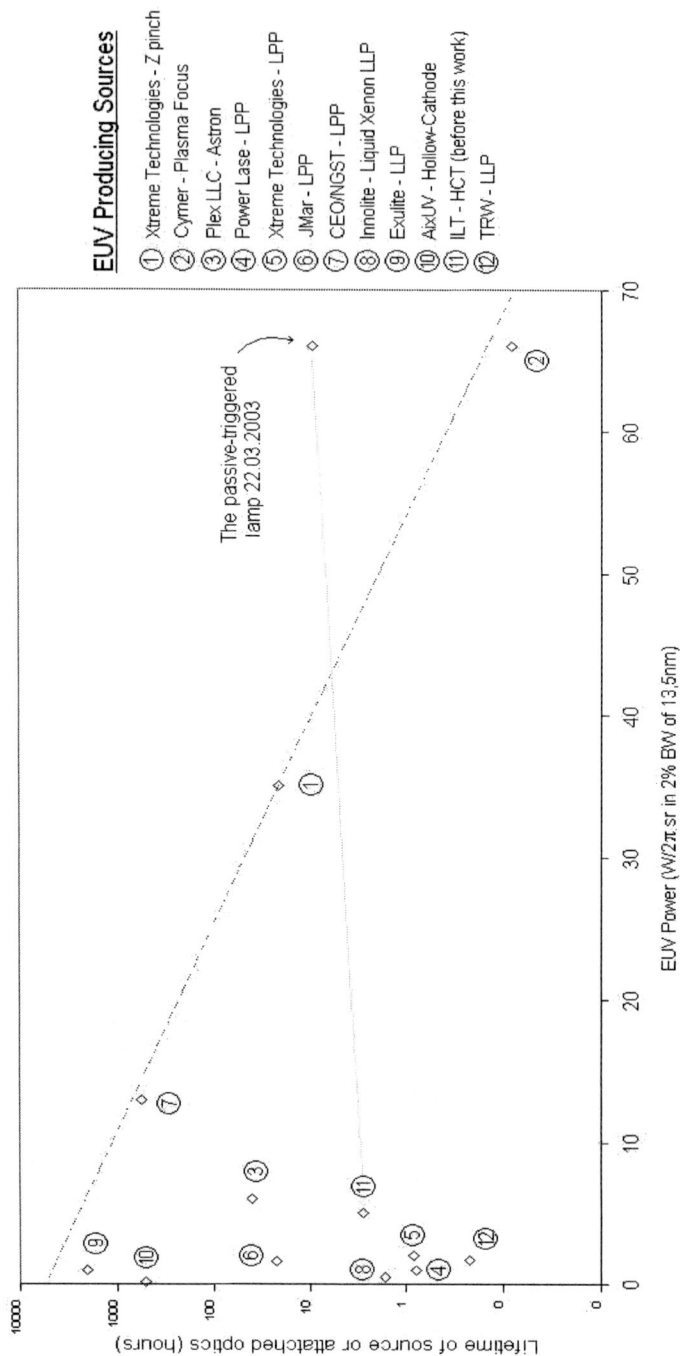

Figure 8.2: Parameter range for available technology, showing the improvement gained by the use of the passive triggered HCT lamp.

Through understanding the ignition process, the HCT lamp has been brought one step closer to meeting the specifications industry demand of an EUV lamp. Figure 8.2 shows the current position of the passive triggered HCT lamp compared with other state of the art technology. Much progress is still required in all relevant parameters, and as the physical limit of xenon is being reached, other radiators, even non-gases, such as tin, need now to be considered, as these are better radiators of 13·5 nm EUV. While some of the concepts of triggering may be transferable, new challenges of how such a system will function await.

9. Appendix

9.1. Paschen's Law

In 1889, Friedrich Paschen published a paper[52] which set out what later became known as Paschen's Law. The law basically states that the breakdown voltage of a gap is a function of the product of gas pressure and the gap separation, usually written as

v = f (p.d),

where p is the pressure and d is the gap separation. More accurately the pressure should be replaced by the number density, N, as it is the number of available gas molecules that counts, i.e., V = f (N.d), which is, of course, affected by the temperature as well as the pressure of the gas through the gas law n/V = p/RT. In our case the remaining ionisation left from the previous breakdown also has a very large influence as the frequency is increased.

For air, and a gap separation on the order of a few millimetres, the breakdown voltage is almost a linear function of the gap length:

$V = 3.10^4.p.d + 1.35$ kV

where d is in centimetres, and p in mbar.[17]

Much research has been done since 1889 to provide a theoretical basis for the law and to develop a greater understanding of the breakdown mechanisms. It should be realised though that there are many factors which have an effect on the breakdown of a gap, such as radiation, dust or surface irregularities. Excessive theoretical analysis might help understand why a gap breaks down, but won't necessarily provide a more accurate value for the breakdown voltage for any given situation.

Paschen's law reflects the Townsend breakdown mechanism in gases, i.e., a cascade of secondary electrons emitted by collisions in the gap. The significant parameter is p.d, the product of the gap separation and the pressure. Typically, the Townsend mechanism (and

Paschen's law) applies to p.d values of less than 1000 mbar.cm, or gaps around a centimetre at atmospheric pressure. Also, some modifications are necessary for highly electronegative gases because of them recombining the secondary electrons very quickly.

In general, an equation for breakdown can be derived, and suitable parameters chosen by fitting to empirical data. The breakdown voltage,

$$V = \frac{B.p.d}{C + \ln(p.d)}$$

where

$$C = \ln\left(\frac{A}{\ln\left(1 + \frac{1}{\gamma}\right)}\right)$$

and γ is the secondary ionisation coefficient, the net number of secondary electrons produced per incident positive ion. It is a function of the pressure. Electronegative gases (SF_6, Freon, O_2 or CO_2) reattach the electrons very quickly, so they have low γ values. If we take values for air [54]

$\gamma = 10^{-2}$

$A = 11\cdot3$ cm^{-1} mbar^{-1}

$B = 274$ Vcm^{-1} mbar^{-1}

then a value of

$C = 0\cdot89$ is found,

giving:

$$V = \frac{365.p.d}{1\cdot47 + \ln(p.d)}$$

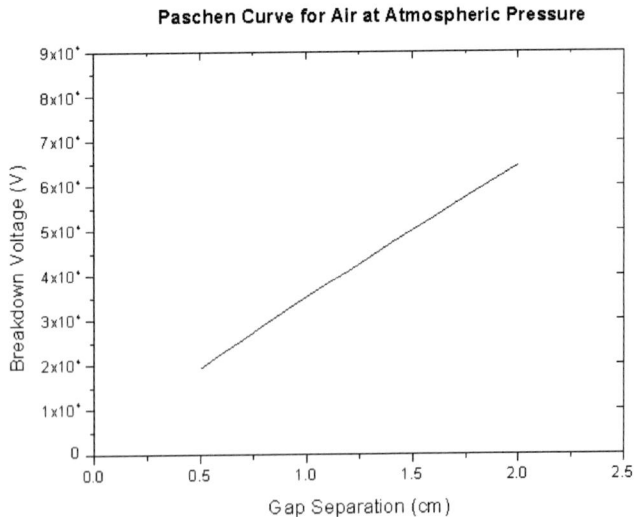

Paschen Curve for Air at Atmospheric Pressure

Figure 9.1: Breakdown voltage of a gap in the cm range for air at atmospheric pressure.

9.2. Additions and Calculations

9.2.1. The Mean Free Path of an Electron in Xe at 15 Pa

If we take a xenon pressure of 15 Pa, then using:

$$PV = nkT$$

the number of molecules per cubic metre is:

$$N = \frac{P}{kT} = \frac{15 Pa}{1 \cdot 38.10^{-23}.300K} = 3 \cdot 6.10^{21} \text{ m}^{-3}$$

The total ionisation cross-section for xenon[41] with an energy of around 3 keV (i.e. the voltage on the main gap) is $\sim 10^{-16}$ cm^2.

Or: $\sigma = 10^{-20}$ m^2

The mean free path, $\lambda = \dfrac{1}{N\sigma} = \dfrac{1}{3 \cdot 6.10^{21}.10^{-20}} = \dfrac{1}{36 \text{ m}^{-1}} = 3 \text{ cm}$

9.2.2. The Maximum speed of a Xe Ion in a 3 kV Field

If given the maximum energy from the field, a singularly charged xenon ion will have a kinetic energy of 3 keV. This corresponds in Joules to:

$$E = 3000 \text{ keV}. \ 1 \cdot 6021.10^{-19} \text{ J.eV}^{-1} = 4 \cdot 81.10^{-16} \text{ J}$$

The atomic mass of xenon is 131·29 g/mol. This corresponds to a mass for one xenon atom of:

$$m = \frac{A}{1000.N_A} = \frac{131 \cdot 29}{1000.6 \cdot 022.10^{23}} = 2 \cdot 18.10^{-25} kg$$

The maximum velocity that the xenon atom can reach is then defined by:

$$V = \sqrt{\frac{2E}{m}} = \sqrt{\frac{2.4 \cdot 81.10^{-16} J}{2 \cdot 18.10^{-25} kg}} = 6 \cdot 6.10^5 m.s^{-1}$$

If the distance from the hollow-cathode to the trigger chamber is 4 cm, the ion will take a time,

$$t = \frac{d}{v} = \frac{0 \cdot 04 \, m}{6 \cdot 6.10^5 \, m.s^{-1}} = 60 \cdot 6 \, ns$$

which corresponds well to the time scale of the lamp system.

9.2.3. The Mean Free Path of a Xenon Ion in Xe at 15 Pa

From above we take the number of molecules per cubic metre, $N = 3 \cdot 6.10^{21} \, m^{-3}$.

The collision cross-section for a Xe Ion at 3 keV is, $\sigma = 5.10^{-19} \, m^2$.

The mean free path, $\lambda = \frac{1}{N\sigma} = \frac{1}{3 \cdot 6.10^{21} . 10^{-19}} = \frac{1}{1800} m = 0 \cdot 6 \, mm$

To obtain a mean free path of 4 cm a pressure of $0 \cdot 2$ Pa would be necessary! This makes it unlikely that ions are triggering the lamp (see 6.2.2).

9.3. Plasma Decay

Plasma can decay in several ways, such as volume recombination of xenon ions with electrons,

$$Xe^+ + e^- \rightarrow Xe + \gamma.$$

Here a neutral atom and a photon, γ are formed. While this collision will take place quite often, the recombination is however unlikely to play such an important role as the photon cannot carry much momentum and an elastic collision is more likely to take place. The time scale required for volume recombination to decay the plasma in the lamp is much larger than the time between the shots[42], represented by [43]

$$\tau = 1 \cdot 14.10^{26} T_e^{4,5} n_e^{-2} \approx 11\,\text{ms},$$

where T_e is the electron temperature (~1 eV) and n_e the electron density (~10^{14} cm^{-3}). A decay time of 11 ms corresponds only to a frequency of 90 Hz, so volume recombination alone cannot be the limiting factor as current repetition rates are much higher than 90 Hz. From this it can be seen why the active trigger with its large volume cannot reach high frequencies with high pressures. Many of the earlier switch designs had large volumes, but ran at low pressures or with lighter gases, such as H_2 or He, which speed up the decay process.

If an ion meets with two electrons (Thompson recombination),

$$Xe^+ + e^- + e^- \;\rightarrow\; Xe + e^-$$

this would allow the remaining energy to be removed, but requiring three particles, is again not so likely.[44] The same applies for two step decay processes, such as

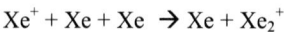

$$Xe^+ + Xe + Xe \;\rightarrow\; Xe + Xe_2^+$$

$$\rightarrow \quad Xe_2^+ + e^- \;\rightarrow\; Xe + Xe$$

The most likely process to take place is believed to be recombination at the walls of the system (the electrodes), here, there is a sufficient supply of electrons and the energy can easily be dissipated to the walls. [42), 45]

$$Xe^+ + e^- + Wall \;\rightarrow\; Xe + Wall$$

Assuming the plasma decay is described by the standard diffusion equation:

$$\frac{\partial n_e}{\partial t} = -\nabla(D \cdot \nabla n_e)$$

with a diffusion coefficient D, assumed here to be independent of the position, for a planar geometry, like between the anode and cathode this can be simplified to a one dimensional expression:

$$\frac{\partial n_e}{\partial t} = -D \frac{\partial^2 n_e}{\partial x^2}$$

Then, with the boundary conditions that the plasma decays at the walls, i.e. the electron density must be zero, a solution to the differential equation can be found to be:

$$n_e(x,t) = n_e^0 \sin\left(\pi \frac{x}{d}\right) e^{-\frac{t}{\tau}}$$

Other physical solutions do exist and could be used to fit the solution to the initial electron density. Here though it will be assumed that the initial density can be written as:

$$n_e(x,0) = n_e^0 \sin\left(\pi \frac{x}{d}\right)$$

which can then be entered back into the original equation giving:

$$n_e^0 \cdot \sin\left(\pi \frac{x}{t}\right) \cdot \frac{-1}{\tau} e^{-\frac{t}{\tau}} = -D \cdot n_e^0 \cdot \left(\frac{\pi}{d}\right)^2 \sin\left(\pi \frac{x}{d}\right) \cdot e^{-\frac{t}{\tau}}$$

and an expression results for the decay time of the plasma, τ:

$$\tau_{planar} = \frac{1}{D}\left(\frac{d}{\pi}\right)^2$$

A similar solution can be found for a cylindrical geometry (hollow-cathode or bore hole) and this gives a Bessel distribution, from which the decay time can be found:

$$\tau_{cylindrical} \approx \frac{1}{D}\left(\frac{R}{2 \cdot 40}\right)^2$$

where R is the radius of the cylinder. To find some numerical values the ambipolar diffusion coefficient is required, defined by:[58]

$$D = D_i \left(1 + \frac{T_e}{T_g} \right)^2$$

where T_e and T_g are the electron and gas temperatures respectively. The ion diffusion coefficient, D_i is defined by [44]

$$D_i = \frac{3}{8} \frac{k_B T_g}{p} \frac{k_B T_g}{M\Omega(T_g)}$$

which is inversely proportional to the pressure, p, the ionic mass, M, and the ion-atom collision integral, Ω. The value of Ω increases slowly with the gas pressure and has typical values of $7 \cdot 5.10^{-16}$ m^{3}s^{-1} for helium or $4 \cdot 8.10^{-16}$ m^{3}s^{-1} for argon. If a value of 3.10^{-16} m^{3}s^{-1} is taken for xenon and for a typical pressure of 20 Pa and an estimated temperature of 1000 K, an ion diffusion coefficient of 5.10^{-2} m^{2}s^{-1}, an ambipolar diffusion coefficient of around $0 \cdot 1$ m^{2}s^{-1} is found, assuming $T_e = T_g$. Results using this calculation are shown in section 4.3.3.

10. Literature List

1   V. Banine, R. Moers, Proc. SPIE, Vol. 4343, 203-214 (2001)

2   G. E. Moore, Electronics, Volume 38, Number 8, April 19, (1965)

3   Aschke, Mengel und Rau, C't Heft 13, Lithographie am Limit, (2003)

4   J.E. Bjorkholm, "EUV Lithography-The Successor to Optical Lithography?", Intel Tech. Journal Q3. (1998)

5   N. R. Pereira and J. Davis, J. Applied Physics. 64 (3), R1 (1988)

6   Technologiestichting STW, "Excimer-laser induced plasma as X-ray source for X-ray projection lithography".

7   M. Bauer et al., "A 20GW-pinch as an ion and X-ray source", proc. 6th IEEE International Pulsed Power Conference, Arlington, Virginia (1987)

8   Proceedings of "Micro- and Nano-engineering", Leuven, Belgium (1998)

9   Vinogradov and Shlyaptsev, "Characteristics of a Plasma X-ray Source (review)", (1987)

10 Presentation by XTREME technologies, "High Power EUV sources for lithography" at the ISMT EUV Source Workshop in Antwerp, Belgium, (2003)

11 Presentation by Cymer, "High Power EUV sources for lithography" at the 3rd International Workshop on EUV Lithography, Matsue, Japan (2001)

12 L. Rapezzi, "Plasma Focus", Web site of the Italian National Agency for new Technologies, Energy and Environment: http://www.enea.it/

13 K. Bergmann et al., "Triggering a radial multichannel pseudospark switch using electrons emitted from a material with high dielectric constant", Appl. Phys. Lett. 71 (14), (1997)

14 M. McGeoch, "High Power EUV Source", EUVL Symposium, Dallas (2002)

15 K. Wiesemann, "Einführung in die Gaselektronik", B.G. Teubner, Stuttgart, (1976)

16 Pankert, Bergmann,Wester, Klein, Neff, Rosier, Seiwert, Smith, Probst, Vaudrevange, Siemons, Apetz, Jonkers, Loeken, Derra, Krücken, Zink, "Hollow Cathode Triggered Plasma Pinch Discharge", SPIE Microlithography, to be published (2004)

17 J. D. Cobine, "Gaseous Conductors: Theory and Engineering Applications", Dover, NY, (1958)

18 J. Christiansen and C. Schultheiss, "Production of high current particle beams by low pressure spark discharges", Z. Phys. 290 35 (1979)

19 M. J. Rhee, "General Information on Pseudospark", http://www.ee.umd.edu/~rhee/pseudospark/, (1998)

20 Boeuf and Pitchford, "Pseudospark Discharges via Computer Simulation", IEEE Trans. Plasma Sci. 19, (1991)

21 J. Christiansen, "The Properties of the Pseudospark", in M. A. Gundersen, "Physics and Applications of Pseudosparks", ISBN: 0-306-43539-X, (1990)

22 Koval, Oks, Schanin, Yu, Kreindel and Gavrilov, "Broad beam electron sources with plasma cathodes", Nuc. Instr. Meth, Phys Res., vol A321 (1992)

23 Kozyrev, Korolev and Shemyakin, "Processes in the cathode region of a low pressure arc discharge", Russian Phys. J. vol. 37, no. 3 (1994)

24 Klimenko et al., "Mechanism for discharge initiation in pseudospark low-pressure spark gaps", Sov. J. Plasma Phys., vol 17, no. 10 (1991)

25 Kozyrev et al., "Process in the breakdown stage of a low pressure discharge and mechanism of discharge initiation in pseudospark switches", J. Appl. Phys., vol 74, no. 9 (1993)

26 Anders, Anders and Gundersen, "Self-Sustained Self-Sputtering: A Possible Mechanism for the Superdense Glow Phase of a Pseudospark", IEEE Trans. on Plasma Sci. Vol. 23, (1995)

27 M. Settler et al., "Firts Experimental Observation of the Ignition of a 'Superdense Glow' before the Glow-to-Arc Transistion in a Pseudospark Discharge", IEEE Trans. on Plasma Sci. Vol. 79, (1996)

28 Krall & Trivelpiece, Elementary Plasma Physics, San Francisco Press, (1986)

29 K. Frank, Yu. Korolev and A. Kuzmichev, "Mechanism for Initiation of Pseudospark Discharge by Ions Injected From the Anode Side", IEEE Trans. on Plasma Sci. Vol. 30, (2002)

30 K. Frank, A. Görtler, F. Heine, D. Hofmann, "Trigger devices for pseudospark switches", IEEE Trans. on Plasma Sci. Vol. 23, (1995)

31 Korolev et al., " Low-Voltage Triggering for a Pseudospark Switch With an Auxiliary Glow Discharge", IEEE Trans. on Plasma Sci. Vol. 29, (2001)

32 Korolev and Frank, "Discharge formation process and glow-to-arc transition in pseudospark switch", IEEE Trans. on Plasma Sci. vol. 27, (1999)

33 G. Mechtersheimer et al., "High repetition rate, fast current rise, pseudo-spark switch.", J. Phys. E: Sci. Instrum. 19, (1986)

34 Yu. Korolev in private communication, Philips EUV GmbH, Aachen (2002-2003)

35 H. Gundel, H. Riege, J. Handerek, and K. Zioutas, Appl. Phys. Lett. 54, 2071, (1989)

36 Kiefer, "Der radiale Mehrkanal-Pseudofunkenschalter als Ersatz für Funkenstrecken", dissertation, Fraunhofer Institut für Laser Technik, (1997)

37 Kirkman, Hartmann and Gundersen, "Flash-lamp-triggered high-powered thyratron-type switch", Appl. Phys. Lett. 52, 613, (1988)

38 Bochkov et al., "Investigation of the Effect of Blocking Potential on the Static Breakdown Voltage and Discharge Initiation in Pseudospark Switches", IEEE Trans. on Plasma Sci. Vol. 23, (1995)

39 Pankert, Bergmann, Klein, Neff, Rosier, Seiwert, Probst, Vaudrevange, Siemons, Bosch, Zink, Krücken, Smith, Apetz, Jonkers, Loeken and Derra, "Physical Properties of the HCT source", SPIE Microlithography, Santa Clara, Proc. SPIE 5037, 15 (2003)

40 Pflüger, "Microwellenanregung für Edelgasionenlaser", Diplomarbeit, RWTH-Aachen (1989)

41 Rapp, D., Englander-Golden, P., J. Chem. Phys. 43, 1464, (1965)

42 Rosier et al., "Frequency Scaling in a hollow-cathode triggered pinch plasma as radiation source in the extreme ultraviolet", (IEEE - to be published)

43 R.Wester and M.Aden, Personal communication, Fraunhofer Institut für Lasertechnik, Aachen (2001-2004)

44 McDaniel, "Collision Phenomena in Ionized Gases", Wiley (1964)

45 J. Jonkers, personal communication, Philips EUV GmbH, Aachen (2001-2004)

46 K. Bergmann, personal communication, Fraunhofer Institut für Lasertechnik, Aachen (2001-2004)

47 Hagstrum, "Auger Ejection of Electrons from Tungsten by Noble Gas Ions", Phys.Rev. vol. 96, 325 - 336 (1954)

48 M. M. Pedović et al, "Electrical breakdown in low pressure gases", J. Phys. D: Appl. Phys. 35, R91-R103 (2002)

49 A. Naweed, "Einfluß der Startelektronen auf den Zündvorgang der Hohlkathode eine Pseudofunkenentladung", Diplomarbeit, RWTH-Aachen (1994)

50 C. Smith, "Emission of X-radiation from a Pseudospark Plasma Discharge", M.Sc. Thesis, University of Birmingham, (1998)

51 J. Aghassi, "Stochastische Heizung in Particle-in-Cell Simulationen von Plasmagasentladungen", Master's Thesis, RWTH-Aachen (2003)

52 F. Paschen, "Über die zum Funkenübergang in Luft, Wasserstoff und Kohlensäure bei verschiedenen Drücken erforderliche Potentialdifferenz", Wied. Ann., 37, 69, (1889)

53 K. Higuchi, T. Tsutomu, M. Itagaki, T. Sato and Y. Abe, "Effects of Electrode Geometry and Gas Pressure on Breakdown Voltage of a Pseudospark Discharge", J. Applied Physics Vol. 35, (1996)

54 Bazelyan, "Spark Discharge", CRC Press, Boca Raton, ISBN: 0-849-32868-3, (1998)

55 R. Bischoff, "The influence of a blocking electrode and fast charging on the working pressure of a Pseudospark discharge in Xenon", DPG-Tagung, Sitzung P.10, Aachen (2003)

56 Sommerer, Pak and Kushner, "Cathode heating mechanisms in pseudospark plasma switches", J. Appl. Phys, 72(8), 3374 (1992)

57 Alberta, Derouard, Ouadoudi, Boeuf and Pitchford, "Space and time dependence of the electric field and plasma induced emission in transient and steady-state hollow cathode discharges", Phys. Rev. E, 2239 (1994)

58 Bittencourt, "Fundementals of Plasma Physics", Pergamon Press, 0-08-033923-9 (1988)

59 J.Bjorkholm, "EUV Lithography-The Successor to Optical Lithography?", Intel Technology Journal Q3, 1998.

60 E.Spiller, Soft X-Ray Optics, pg. 185, SPIE, Bellingham, WA, 1994.

Acknowledgement

The work presented here took place at the Fraunhofer-Institut für Lasertechnik and the Lehrstuhl für Lasertechnik of the Rheinisch-Westfälischen Technischen Hochschule in Aachen, together with close collaboration with Philips Extreme UV GmbH. I would like to thank all my colleagues of these three institutions for the pleasant working environment and for their co-operation.

My particular thanks go to Prof. Dr. rer. nat. Reinhart Poprawe M.A. for making it possible to carry out this work, Prof. Dr. rer. nat Kurt Kugeler for accepting the position of co-referee, Dr. Willi Neff, Dr. Joseph Pankert and Dr. Klaus Bergmann for their trust in me and many helpful discussions, Dr. Michael Loeken, Dipl. Ing Sven Probst, Dipl. Ing Guido Siemons, Dr. Jeroen Jonkers, Dr. Juergen Klein und Mr. Oliver Zitzen for their intensive experimental co-operation and support in the laboratory, Dr. Thomas Kruecken, Dr. Mirko Aden, und Dr. Rolf Wester for answering many a theoretical question, Dr. Oliver Rosier, Dr. Rolf Apetz, Dipl. Ing. Dominik Vaudrevange and Dipl. Ing. Stefan Seiwert for their support and co-operation, as well as all other members of the ILT Plasma Technology group and Philips Extreme UV for their friendship and support, my family for their support during all my studies and my doctorate, as well as all other people not mentioned here who have contributed to the success of this work.

I have now finished *school* and have a *real* job!

Curriculum Vitae

24th October 1974 Born in Sheffield, England

1986 - 1991   Old Hall Comprehensive School
Rotherham, England

1991 - 1993   Thomas Rotherham College

1993   A'levels
WH.BACON prize for Physics

1993 - 1996   Imperial College of Science
Technology and Medicine
University of London

1996   Bachelors Degree (Physics)
Associate of the Royal College of Science

1996 - 1997   RWTH-Aachen/Fraunhofer Institut für Lasertchnik

1997 - 1998   University of Birmingham, England

1998   Masters Degree (Applied Radiation Physics)

1998 - 2001   RWTH-Aachen/Research Centre Jülich, Germany

2001 - 2004   Doctorand at the Fraunhofer Institut für Lasertchnik/ RWTH-Aachen

2004 -   Patent Examiner at the European Patent Office in Munich